Welcome to *AMS Climate Studies*!

You are about to explore the complexity and wonder of Earth's climate system. This ***Climate Studies Investigations Manual*** is designed to introduce you to tools that enable you to explore, analyze, and interpret the workings of Earth's climate.

This ***Investigations Manual*** is composed of self-contained investigations. These learning experiences draw from actual climate observations and events to assist students in achieving their stated objectives. The investigations continually build on previous learning experiences to help the learner form a comprehensive understanding of the Earth's climate system – your environment.

Additionally, one case study of a current or recent climatic situation is delivered via the course website in real time each week during fall and spring semesters in a schedule aligned with the ***Investigation Manual***'s table of contents (for Investigations 1A through 12B). These "Current Climate Studies" appear on the course website by noon, Eastern Time, on Monday for optional use as determined by the course instructor. Current Climate Studies accumulate each semester and remain available via a website archive. Studies expanding on ***Manual*** Investigations 13A through 15B are posted to the website at the beginning of each fall semester and are available throughout the year.

Getting Started:

1. Your course instructor will provide you with the specific requirements of the course in which you are enrolled.

2. The ***Climate Studies*** course website login address is:
   http://www.ametsoc.org/amsedu/login.cfm
   (an alternate if necessary is: *http://amsedu.ametsoc.org/amsedu/login.cfm*).
   Record this address, add this address to your list of bookmarks or favorites for future retrievals.

3. When the page comes up, type the login ID and password <u>provided by your instructor</u> when prompted for full access to the contents of the page.

   Login ID: _____
   Password: _____

4. Explore the course website, noting its organization and the kinds of information provided. Throughout the year, 7 days a week, 24 hours a day, the climatic products displayed are the latest available. You will learn to interpret and apply many of these products via the ***AMS Climate Studies*** investigations.

5. Complete ***Investigation Manual*** activities and other course requirements, including use of Current Climate Studies, as directed by your instructor.

6. **Keep Current!** Keep up with the climate news and your climate studies. The climate affects all of us, everywhere, watch it in action. Visit the ***AMS Climate Studies*** website as frequently as you can.

AMS CLIMATE STUDIES
INTRODUCTION TO CLIMATE SCIENCE

INVESTIGATIONS MANUAL

EDITION 3

The American Meteorological Society Education Program

The American Meteorological Society (AMS), founded in 1919, is a **scientific and professional** society. Interdisciplinary in its scope, the Society actively promotes the development and dissemination of information on the atmospheric and related oceanic and hydrologic sciences. AMS has more than 14,000 professional members from more than 100 countries and over 175 corporate and institutional members representing 40 countries.

The Education Program is the initiative of the American Meteorological Society fostering the teaching of the atmospheric and related oceanic and hydrologic sciences at the precollege level and in community college, college and university programs. It is a unique partnership between scientists and educators at all levels with the ultimate goals of (1) attracting young people to further studies in science, mathematics and technology, and (2) promoting public scientific literacy. This is done via the development and dissemination of scientifically authentic, up-to-date, and instructionally sound learning and resource materials for teachers and students.

AMS Climate Studies, the newest component of the AMS education initiative, is an introductory undergraduate Climate Science course offered partially via the Internet in partnership with college and university faculty. **AMS Climate Studies** provides students with a comprehensive study of the principles of Climate Science while simultaneously providing classroom and laboratory applications focused on the rapidly evolving interdisciplinary field of Climate Science.

Developmental work for **AMS Climate Studies** and the companion **DataStreme: Earth's Climate System** was supported by the National Aeronautics and Space Administration under Grants Number NNX09AP58G and NNX08AN53G. Any opinions, findings, and conclusions or recommendations expressed in this material are those of the author and do not necessarily reflect the views of the National Aeronautics and Space Administration.

Climate Studies Investigations Manual 3rd Edition
ISBN-10: 1-935704-99-0
ISBN-13: 978-1-935704-99-7
Copyright © 2012 by the American Meteorological Society

All rights reserved. No part of this publication may be reproduced, stored in a retrieval system, or transmitted, in any form or by any means, electronic, mechanical, photocopying, recording or otherwise, without the prior written permission of the publisher.

Published by the American Meteorological Society
45 Beacon Street, Boston, MA 02108

AMS Climate Studies Investigations

1A   MODERN CLIMATE SCIENCE
Defines "climate", introduces Earth's climate system and investigates the AMS Climate paradigm

1B   FOLLOW THE ENERGY! EARTH'S DYNAMIC CLIMATE SYSTEM
Presents a simplified model of Earth's energy system to receive and emit radiation

2A   CLIMATE SCIENCE FROM AN EMPIRICAL PERSPECTIVE
Examines traditional sources of climatological data

2B   CLIMATE VARIABILITY AND CHANGE
Investigates a unique case of one possible climate change cause

3A   SOLAR ENERGY AND EARTH'S CLIMATE SYSTEM
Contrasts the global distribution of incoming solar energy at various latitudes

3B   ATMOSPHERIC CO_2, INFRARED RADIATION, AND CLIMATE CHANGE
Examines the absorption of infrared radiation by carbon dioxide

4A   WATER, HEAT, AND HEAT TRANSFER
Compares the heat storage of water to other common substances

4B   WATER AND HEAT STORAGE AT THE EARTH'S SURFACE
Contrasts the heat storage in water and soil locations with its implications for Earth's climate system

5A   GLOBAL WATER CYCLE
Investigates remote sensing of Earth's global water cycle

5B   WATER VAPOR FLUX AND TOPOGRAPHICAL RELIEF
Examines the effect of Earth's topography on precipitation patterns

6A   GLOBAL ATMOSPHERIC CIRCULATION
Compares the general circulation of Earth's atmosphere from theory to practice

6B   GLOBAL ATMOSPHERIC CIRCULATION – ROSSBY WAVES
Investigates midlatitude tropospheric motions of Rossby waves

7A SYNOPTIC-SCALE ATMOSPHERIC CIRCULATION – HIGH AND LOW PRESSURE SYSTEMS
Describes synoptic scale weather systems

7B SYNOPTIC-SCALE ATMOSPHERIC CIRCULATION – WAVE CYCLONES AND STORM TRACKS
Investigates development and movement of synoptic scale weather systems

8A CLIMATE AND AIR/SEA INTERACTIONS – INTER-ANNUAL TO DECADAL CLIMATE VARIABILITY
Considers variability in climate from evolving air-sea interactions

8B COASTAL UPWELLING AND COASTAL CLIMATES
Examines coastal climates that result from oceanic upwelling

9A PETM: A POSSIBLE ANALOG TO MODERN CLIMATE CHANGE
Investigates an historic analog climate state to rapid change mechanisms

9B METHANE HYDRATES: MAJOR IMPLICATIONS FOR CLIMATE
Demonstrates one climate system component that may trigger rapid system change

10A CLIMATE AND CLIMATE VARIABILITY FROM THE INSTRUMENTAL RECORD
Explores climate variability evidenced in the empirical record

10B RICE GROWING AND CLIMATE CHANGE
Displays an example of climate variability or change and its possible impact on agriculture

11A VOLCANISM AND CLIMATE VARIABILITY
Explores one natural forcing mechanism determined from the paleoclimate record

11B SNOW AND ICE ALBEDO FEEDBACK IN EARTH'S CLIMATE SYSTEM
Examines an important feedback mechanism in the climate system and its evidence

12A CLIMATE CHANGE AND RADIATIVE FORCING
Compares radiative forcings of the climate system by various natural and human mechanisms

12B THE OCEAN IN EARTH'S CLIMATE SYSTEM
Examines evidence of climate change in Earth's largest energy reservoir – the ocean

13A **VISUALIZING CLIMATE**
Examines traditional displays of empirical climate data and related classification schemes

13B **CLIMATE VARIABILITY AND SHORT-TERM FORECASTING**
Looks at short-term climate predictions based on climate variability

14A **CLIMATE MITIGATION AND ADAPTATION STRATEGIES**
Examines climate mitigation and adaptation strategies for the future

14B **GEOENGINEERING THE CLIMATE**
Considers global geoengineering strategies and feasibility

15A **CLIMATE MITIGATION THROUGH CARBON EMISSION CAP-AND-TRADE**
Investigates the Cap-and-Trade Process for mitigation of climate change

15B **CARBON DIOXIDE EMISSIONS, CARBON FOOTPRINTS, AND PUBLIC POLICY**
Explores the lifetime of atmospheric carbon dioxide and carbon "footprints" as well as public opinion on climate change

MODERN CLIMATE SCIENCE

Driving Question: *What is Earth's climate system and what are the empirical and dynamic definitions of climate?*

Educational Outcomes: To identify some of the many reasons for studying Earth's climate system. To learn more about the workings of Earth's climate system and become more aware of the significance of climate, climate variability, and climate change for our well being wherever we live.

Objectives: *AMS Climate Studies* is an innovative study of Earth's climate system that promises to deliver new understandings and insights into the role of climate in our individual lives and the broader society. The *AMS Climate Paradigm* presented in this Investigation employs an Earth system science approach.

After completing this investigation, you should be able to:

- Describe Earth's climate system and its interacting components.
- Describe, compare and contrast the complementary empirical and dynamic definitions of climate.
- Explain the *AMS Climate Paradigm*.

An Earth System Approach:

AMS Climate Studies employs an Earth system perspective. A view of the Earth system as seen from space is presented in **Figure 1**. The image shown is a visible light full-disk view from a U.S. weather satellite positioned about 36,000 km (22,300 mi) above the equator in South America at 75 degrees W longitude. The satellite remains at that location relative to Earth's surface because it makes a full revolution around the planet as Earth makes one rotation in the same direction. Being geostationary, the satellite provides a continuous view of the same underlying surface. Successive images from this vantage point provide animations of whatever can be seen moving across Earth's surface, including the boundaries, called terminators, which separate the illuminated day side and dark night side of our planet.

Examine Figure 1, noting the outlines of land masses. The center of the disk is the point on Earth directly under the satellite from which this image was acquired. Place a dot on the image to represent this sub-satellite point and draw a horizontal line, representing the equator, through the point and extended to the edges of the Earth disk. Approximately one-third of Earth's surface can be seen from the satellite.

1. Figure 1 is a view of the Earth system with the edge of the disk marking the boundary between Earth and the rest of the universe. It is evident from the sharpness of the edge between Earth and space that the atmosphere must be a thin layer compared to Earth's diameter. Since the full disk appears sunlit in this visible image, the local time at the sub-solar point must be near **[(*noon*)(*sunset*)(*sunrise*)]**.

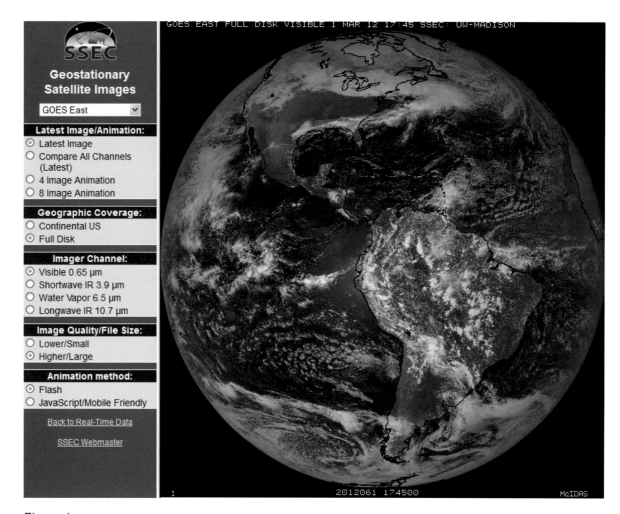

Figure 1.
Visible image of Earth from NOAA GOES East satellite at 1745 UTC on 1 March 2012. The time was 12:45 pm EST, 11:45 am CST, 10:45 am MST, 9:45 am PST.

2. Compare land and ocean surfaces in this view. As would be seen from other vantage points in space as well, Earth's surface is [(*more*)(*less*)] water than land.

Figure 1 is a static view of Earth's climate system. For a view of it in motion, go to: *http://www.ssec.wisc.edu/data/geo/index.php?satellite=east&channel=vis&coverage=fd &file =jpg&imgoranim=8&anim_method=flash* , or *http://www.ssec.wisc.edu/data/geo/ index.php? satellite=east&channel=vis&coverage=fd&file=jpg&imgoranim=8&anim_ method=jsani*

Short Cuts! Please note that all web addresses appearing in these investigations are available on the course website by clicking on "Investigations Manual Web Addresses". Under the particular Investigation heading, click on the link shown. The website above also can be accessed via QR coding presented at the right.

QR Code

1A - 3

You are viewing the University of Wisconsin Space Science and Engineering Center (SSEC) website from which Figure 1 was acquired. The animation that appears is composed of eight recent full-disk images from the GOES East satellite, most acquired at three-hour intervals with the latest being only a few hours old. View the animation that essentially covers one day as it repeats through several cycles. To look at individual images or to slow down the animation, use the control bar above the image. First, click on "stop". Then click successively on the step-forward (>) button while noting the progression of day and night on Earth's surface as the rotating planet intercepts the radiant energy from the distant Sun.

3. Half of Earth's surface is in sunlight and half in darkness. The sunlit portion in the image shows the part of Earth in the satellite's field of view that is receiving energy from outside the Earth system. The animation shows that the solar energy coming into the Earth system at any location is [(*continuous, constantly illuminating the surface*)(*received in pulses, alternating between periods of sunlight and no sunlight*)].

4. The time of each image is printed across the top, after the date. Stop the animation at the 1745 UTC image, the same time of day as the Figure 1 image. Compare it with Figure 1. The major observable differences in the two images arise from the Earth system's [(*land surfaces*)(*ocean surfaces*)(*atmosphere*)].

Because these images are visible light images (essentially conventional black and white photographs), features are distinguished by the variation and quality of reflected sunlight. Generally, the brighter (whiter) the feature, the greater the reflection of solar radiation directly back to space. Conversely, darker areas indicate greater absorption of the incoming solar energy.

5. Generally, the image shows that [(*land surfaces*)(*water surfaces*)(*cloud tops*)] are places where the greatest amount of incoming solar energy is absorbed into the Earth system.

On the SSEC Geostationary Satellite Images browser menu to the left, click on the Imager Channel "Longwave IR 10.7 µm" button. Here you are viewing images of "heat" radiation emitted by the Earth system out to space. In these IR images, the darker areas represent those places where outgoing heat radiation to space is greater, and lighter areas denote less outgoing heat radiation. Essentially, these are images of temperature. The darker the shading, the higher the temperature of the surface from which the radiation is being emitted and the greater the rate at which heat energy is being lost to space.

6. Comparison of the IR animation with the visible light animation shows that the Earth system emits IR to space [(*continuously*)(*only on the night side of Earth*)].

7. Step through the IR animation for several cycles and look for broad, essentially cloud-free places where shading changes most, that is, they alternate between dark shading (meaning they reach relatively high temperatures) and light shading (meaning cooler temperatures) over the period of a day. These locations are [(*land*)(*water*)] surfaces.

Climate Studies: Investigations Manual 3rd Edition

8. Stop the IR animation on the image with places shaded darkest and note the time of the image. Switch to the visible imagery by going to the browser menu and stopping the animation at the same time. The comparison shows that the highest surface temperatures occur within a few hours of local [(*__midnight__*)(*__sunrise__*)(*__mid-day__*)].

In summary, you have been introduced to the Earth system, the receipt of sunlight into the Earth system from space (incoming energy), and the emission of IR (heat) from Earth to space (outgoing energy).

That part of our planet (including the atmosphere, ocean, land, biosphere and cryosphere) subjected to solar energy flowing into, through, and out of the Earth system, is ***Earth's climate system***.

Weather, Climate and Climate Change:

Fundamental to an understanding of weather, climate and climate change, is the recognition that the Earth's climate system is a complex system of energy flow, as alluded to by animations of visible and IR full-disk views of Earth. The observable impacts of the energy flows (and the associated mass flows) are embodied in the descriptions of weather and climate.

Weather is concerned with the state of (i.e., conditions in) the atmosphere and at Earth's surface at particular places and times. Weather, fair or stormy, is not arbitrary or capricious. Both its persistence and its variability are determined by energy and mass flows through the Earth system.

Climate is commonly thought of as a synthesis of actual weather conditions at the same locality over some specified period of time, as well as descriptions of weather variability and extremes over the entire period of record at that location. Climate so defined can be called ***empirical***, i.e., dependent on evidence or consequences that are observable by the senses. It is empirical as it is based on the descriptions of weather observations in terms of the statistical averages and variability of quantities such as temperature, precipitation and wind over periods of several decades (typically the three most recent decades).

Climate can also be specified from a ***dynamic*** perspective of the Earth environment as a system. The definition of Earth's climate system must encompass the hydrosphere including the ocean, the land and its features, the biosphere, and the cryosphere including land ice and snow cover, which increasingly interact with the atmosphere as the time period considered increases. While the transitory character of weather results from it being primarily an atmospheric phenomenon, **climate exhibits persistence arising from it being essentially an Earth system phenomenon**.

From the dynamic perspective, climate is ultimately the story of solar energy intercepted by Earth being absorbed, scattered, reflected, stored, transformed, put to work, and eventually emitted back to space as infrared radiation. As energy flows through the Earth system, it determines and bounds the broad array of conditions that blend into a slowly varying persistent state over time at any particular location within the system.

Whereas the empirical approach allows us to construct descriptions of climate, the dynamic approach enables us to seek explanations for climate. Each has its powerful applications. In combination, the two approaches enable us to explain, model and predict climate and climate change. In this course we will treat climate from the two complementary perspectives.

9. In its definition of **climate**, the AMS *Glossary of Meteorology, 2nd. ed., 2000*, states that climate "… *is typically characterized in terms of suitable averages of the climate system over periods of a month or more, taking into consideration the variability in time of these average quantities.*" This definition is derived from a(n) **[(*empirical*)(*dynamic*)]** perspective.

10. The *AMS Glossary*'s definition continues: "… *the concept of climate has broadened and evolved in recent decades in response to the increased understanding of the underlying processes that determine climate and its variability.*" This expanded definition of climate is based on a(n) **[(*dynamic*)(*empirical*)]** perspective.

11. Local climatic data, including records of observed temperature, precipitation, humidity, and wind, are examples of **[(*dynamically*)(*empirically*)]** derived information.

12. The determination of actual **climate change**, also from the *AMS Glossary*, ("*any systematic change in the long-term statistics of climate elements sustained over several decades or longer*") is based primarily on evidence provided from a(n) **[(*dynamic*)(*empirical*)]** perspective.

13. Also from the *AMS Glossary,* "*Climate change may be due to natural external forcings, such as changes in solar emission or slow changes in Earth's orbital elements; natural internal processes of the climate system; or anthropogenic (human caused) forcing.*" This is a statement derived from a(n) **[(*dynamic*)(*empirical*)]** perspective.

14. Scientific predictions of such an altered state of the climate (i.e., climate change) must be based on treating Earth's climate system from a(n) **[(*dynamic*)(*empirical*)]** perspective.

Earth's Climate System (ECS) **Paradigm:**

Utilizing a planetary-scale Earth system perspective, this course explores Earth's climate system. In pursuing this approach, understanding is guided and unified by a special paradigm:

AMS Climate Paradigm

The climate system determines Earth's climate as the result of mutual interactions among the atmosphere, hydrosphere, cryosphere, geosphere, and biosphere and responses to external influences from space. As the composite of prevailing weather patterns, climate's complete description includes both the average state of the atmosphere and its variations. Climate can be explained primarily in terms of the complex redistribution of heat energy and matter by Earth's coupled atmosphere/ocean system. It is governed by the interaction of many factors, causing climate to differ from one place to another and to vary on time scales from seasons to millennia. The range of climate, including extremes, places limitations on living things and a region's habitability.

Climate is inherently variable and now appears to be changing at rates unprecedented in recent Earth history. Human activities, especially those that alter the composition of the atmosphere or characteristics of Earth's surface, play an increasingly important role in the climate system. Rapid climate changes, natural or human-caused, heighten the vulnerabilities of societies and ecosystems, impacting biological systems, water resources, food production, energy demand, human health, and national security. These vulnerabilities are global to local in scale, and call for increased understanding and surveillance of the climate system and its sensitivity to imposed changes. Scientific research focusing on key climate processes, expanded monitoring, and improved modeling capabilities are increasing our ability to predict the future climate. Although incomplete, our current understanding of the climate system and the far-reaching risks associated with climate change call for the immediate preparation and implementation of strategies for sustainable development and long-term stewardship of Earth.

15. It is implied in the *AMS Climate Paradigm* that components of Earth's climate system (e.g., atmosphere, hydrosphere, cryosphere, geosphere, and biosphere) interact in a(n) [(*random*)(*orderly*)] way as described by natural laws.

16. This interaction of Earth system components through natural laws would imply a(n) [(*dynamic*)(*empirical*)] perspective for climate studies.

17. The ocean as an Earth system component and player in atmosphere/ocean energy and mass distributions suggest it is a [(*minor*)(*major*)] part of biogeochemical cycles (e.g., water cycle, carbon cycle) operating in the Earth system.

18. According to the *AMS Climate Paradigm*, our understanding of Earth's climate system is incomplete. Nonetheless, it states that the risks associated with climate change call for the development and implementation of [(*sustainable development strategies*)(*long-term stewardship of our Earthly environment*)(*both of these*)].

Summary:

In this course we will investigate climate, climate variability, and climate change through complementary **empirical** and **dynamic** approaches guided by the *AMS Climate Paradigm*.

Please note that Figure 1 and all other Investigations Manual images are also available on the course website. To view these images, click on the "Investigations Manual Images" link on the website, go to the row containing the appropriate investigation name, and then select the appropriate figure within that row. For example, to view Figure 1 online, go to the row labeled "1A" and then select "Fig. 1".

FOLLOW THE ENERGY! EARTH'S DYNAMIC CLIMATE SYSTEM

Driving Question: *How does energy enter, flow through, and exit Earth's climate system?*

Educational Outcomes: To consider Earth's climate as an energy-driven physical system. To investigate fundamental concepts embodied in considering Earth's climate from a dynamic perspective and through the use of models.

Objectives: The flow of energy from space to Earth and from Earth to space set the stage for climate, climate variability, and climate change. After completing this investigation, you should be able to describe fundamental understandings concerning:

- The global-scale flow of energy between Earth and space.
- The impact of the atmosphere on the flow of energy to space.
- The effect of incoming solar radiation on Earth's energy budget.
- The likely effects of energy concentrations and flows on Earth system temperatures.

Earth's Dynamic Climate System

Earth's climate is a dynamic energy-driven system. **The radiant energy received from space and that lost to space on a global basis determine whether or not Earth is in a steady-state condition, cooling, or warming.** An unchanging balance between incoming and outgoing radiation produces a steady-state and stable climate. Lack of a balance between incoming and outgoing radiation implies a net loss or gain of radiant energy to Earth's climate system. **One result of such an energy imbalance is climate change.**

Earth's climate evolves under the influence of its own internal dynamics and because of changes in external factors that perturb the planet's energy balance with surrounding space. The **three fundamental ways in which this energy balance can be disturbed** are by changes in the amount of:

1. solar radiation reaching the Earth system;
2. incoming solar radiation that is absorbed by the Earth system; and,
3. infrared (heat) radiation emitted by the Earth system to space.

Solar radiation intercepted and absorbed by Earth drives our planet's climate system. Earth responds to this acquired energy through the emission of long-wave infrared (heat) radiation as its climate system adjusts towards achieving global radiative equilibrium with space. Because the amount of solar energy intercepted by Earth can be determined with great accuracy by instruments onboard Earth-orbiting satellites, the stage is set for the development of climate models with the potential of predicting future states of Earth's global-scale climate system. In addition to predicting future climate, these climate models can be manipulated quantitatively (e.g., changing the atmospheric concentrations of heat-trapping gases) to provide insight into the probable consequences of various human activities (e.g., combustion of fossil fuels, land clearing).

In this course, the *AMS Conceptual Energy Model* (**AMS CEM**) will be employed to investigate basic concepts underlying the global-scale flows of energy to and from Earth.

This investigation explores energy flow in a highly simplified representation of an imaginary planet and the space environment above it. The purpose is to provide insight into the impacts of physical processes that operate in the real world. This investigation follows the flow of energy as it enters, resides in, and exits a planetary system model, as shown in **Figure 1**. As seen in Figure 1 (a), short-wave solar energy is intercepted by the planet and absorbed at its surface. In Figure 1 (b), the solar-heated surface emits long-wave infrared radiation upwards. In the absence of an atmosphere, the upward-directed radiation would immediately be lost to space. With a clear, cloud-free atmosphere added to the planet, as in Figure 1 (c), some of the upward-directed radiating energy would be absorbed by molecules of heat-trapping greenhouse gases (primarily H_2O and CO_2). **The absorbed energy subsequently radiates from the molecules to their surroundings randomly in all directions, with essentially half of the emissions exhibiting a downward component and half an upward component.** While upward emissions can escape to space, the energy directed downward can return to the planet's surface and add to the amount of energy contained in the planetary climate system.

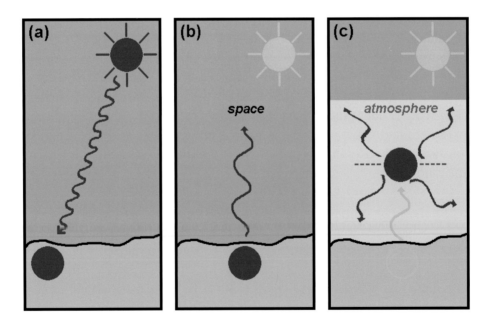

Figure 1.
(a) Sunlight heats the surface of the planet. (b) In absence of an atmosphere, the surface emits infrared radiation to space. (c) If there is an atmosphere, greenhouse gases absorb infrared radiation emitted from the planet's surface and then radiate the energy in all directions, with half directed downward and half upward.

Starting the AMS CEM Investigation:

The *AMS Conceptual Energy Model* (**AMS CEM**) is a computer simulation designed to enable you to track the paths that units of energy might follow as they enter, move through, and exit an imaginary planetary system according to simple rules applied to different scenarios. For simplicity, consider *units of energy* to be equivalent bundles or parcels of energy. To access the interactive model, go to the course website and in the **"Extras"** section, click on AMS Conceptual Energy Model. Then click on *Run the AMS CEM*.

As shown in **Figure 2,** the AMS CEM is presented as a landscape view of a planetary surface, with the Sun depicted in the upper right corner. The AMS CEM is manipulated by choosing different combinations of conditions via windows along the top of the view. Once the conditions have been set, click **Run** to activate the AMS CEM.

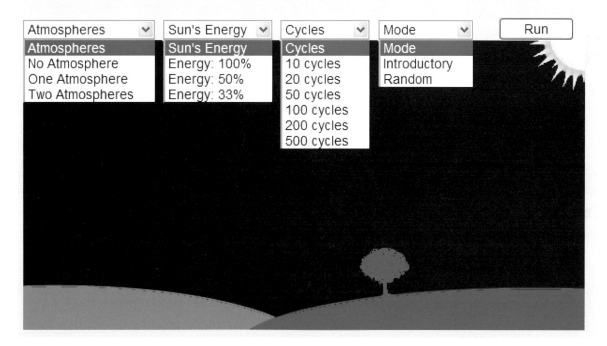

Figure 2.
Landscape view of AMS CEM showing possible choices or settings to conduct model runs.

Become acquainted with the AMS CEM. Start by selecting "One Atmosphere" under **Atmospheres** and "Energy: 100%" under **Sun's Energy** (denoting the arrival of a fresh unit of energy from the Sun during each cycle of play). Select "10 cycles" under **Cycles** so a model run will be composed of 10 cycles of play. Select "Introductory" under **Mode**. Finally, click on **Run**. Because the model is in the Introductory mode, you can observe the same run repeatedly without the cycle patterns changing. You can also stop a run at any time by clicking on **Pause** in the Run window, and then continue the run by clicking on **Resume** in the same window. Note that each model run starts with one unit of energy already at the planet's surface. An atmosphere, if present, does not absorb any of the incoming sunlight passing through it.

1. Repeating or stepping through the run specified above as many times as necessary, <u>follow the first energy unit that originated from the Sun</u>. As it arrives at the planet's surface, the yellow energy unit changes to **[(*green*)(*blue*)(*red*)]**. This signifies its transition from sunlight to heat energy as it is absorbed into the planet's climate system.

In the AMS CEM, a ***cycle of play*** refers to a sequence of moves in which every energy unit in the planet system is subjected to one vertical move. A model ***run*** is composed of a specified number of cycles of play (i.e., 10, 50, 200). For example, a 10-cycle run of the model indicates that whatever energy there is in the planetary climate system at the beginning of <u>each of the 10 cycles of play</u> is subjected to one vertical-motion play during the individual cycle.

Once an energy unit is in the planet system, the **two rules to be followed** as it flows through the planet system during each run of the AMS CEM are:

Rule 1. During each cycle of play, any energy unit at the planet's surface will have an equal chance of staying at the planet's surface or moving upward.

Rule 2. During each cycle, any energy unit in the atmosphere will have an equal chance of moving downward or upward.

These rules are primarily based on the fact that regardless the direction an energy unit comes from when it is absorbed by an atmospheric molecule (i.e., CO_2, H_2O), the energy emitted from the gas molecule can be in any direction. Half the emitted radiation will have a downward component, and half an upward component.

2. Once an energy unit has been absorbed into the planet system, it continues to play every remaining cycle in the run until it is either lost to space or is retained somewhere in the planet system. <u>Continue to follow the first energy unit that arrived from the Sun</u> by replaying the run as many times as you wish, or, by stepping through the run by alternately clicking on **Pause** and **Resume**. In the cycle immediately following its absorption at the planet's surface, the energy unit being tracked **[(*stays at the planet's surface*)(*moves up to the atmosphere*)]**.

3. In its next play, the same energy unit **[(*moves up to space*)(*moves down to planet's surface*)]**.

4. Follow the same energy through the subsequent cycles of play. By the end of the 10-cycle run, it **[(*remains in the planet system*)(*was lost to space*)]**.

5. Next, follow the second energy unit to arrive from the Sun. After the cycle <u>following</u> its arrival from the Sun, the second energy unit ends up **[(*at planet's surface*)(*in the atmosphere*)(*in space*)]**.

6. The planet's climate system in the AMS CEM includes the planet's surface and any existing atmosphere. This ***Planet with an Atmosphere*** computer simulation, as with all AMS CEM simulations, starts with one energy unit in the planet system at the planet's surface. After its 10 cycles of play, this particular run shows the planet system (surface and atmosphere) ending up with [(*1*)(*2*)(*4*)(*6*)] units of energy.

Running this and other simulations in the Introductory Mode always produces the same results for the individual simulation. This is because in the Introductory Mode all energy unit movements are determined by the same set of random numbers essentially "frozen" for the purposes of demonstrating how the model works. Random numbers are employed in AMS CEM to assure that energy-unit movements are determined purely by chance. [In the Random Mode a unique sequence of random numbers is generated with every run, so it is extremely unlikely any two runs can be exactly alike and no run can be repeated.]

7. Modify the Mode setting for the AMS CEM simulation you have been examining (One Atmosphere, Energy: 100%, 10 cycles) by clicking on "Random" under the Mode heading. Now click on the Run button, and watch the model go through its 10-cycle run. Play the new simulation several times, looking for similarities and/or differences. With the random setting, different runs of the model produce [(*different*)(*the same*)] results.

The AMS CEM allows you to investigate numerous questions, such as what impact does an atmosphere have on the amount of energy residing in the system. You can explore this question by modifying the settings of the AMS CEM. Select "No Atmosphere". The other settings remain: Energy: 100%, 10 cycles, Random mode.

8. You have now changed the AMS CEM to evaluate a computer simulation of a ***Planet with no Atmosphere***. Click on the Run button and watch the model go through its 10-cycle run. Repeat several times. Comparison of several runs of the simulations with and without an atmosphere, reveals the generalization that more energy is retained in the planet system that [(*has*)(*does not have*)] an atmosphere.

9. Stated another way, comparing the two simulations (with and without an atmosphere) shows that the addition of an atmosphere, containing energy-absorbing molecules, causes the amount of energy in the planet's climate system to [(*increase*)(*remain the same*)(*decrease*)].

Now change the AMS CEM setting to: One Atmosphere, Energy: 100%, 10 cycles, and Introductory mode. Click on the Run button to review the 10-cycle run. Then, sequentially, choose and make "20", "50", and "100" cycle runs. Since the model is running in the Introductory mode, each subsequent higher-cycle run embodies the previous lower-cycle runs. Note that the model speeds up as the number of cycles in a run increases. This is primarily done as a time-saving device when operating the AMS CEM.

Set the model to 200 cycles and click on the Run button. While it is running, note the curves being drawn on the graph directly below the landscape view. This part of the model is reporting (blue curve) the number of energy units in the planet system cycle-by-cycle as the run progresses. It is also reporting a five-cycle *running average* (green curve). Running averages are commonly calculated in climate science to even out short-term variations and reveal trends. They are calculated at the end of each cycle by adding the most recent observed value and dropping the oldest one. This averaging technique is especially useful in the environmental sciences as new observational data are collected.

10. Directly above the graph, the model reports that for the 200-cycle run, the mean (average) number of energy units in the planet system after each cycle was [(*3.0*)(*4.7*)(*6.6*)(*8.2*)].

The model starts a run with one energy unit in the planet system and the arrival of one energy unit from the Sun. An initial "*spin up*" of the model occurs over ten to twenty cycles before it appears to suggest the model has achieved a relatively stable condition. Although the numbers of energy units in the planetary system can vary considerably over several cycles, the long-term trend shows little evidence of either increasing or decreasing.

11. After an initial "spin up" of the model, the number of energy units in the planet system during the Introductory-mode 200-cycle run ranged between [(*0*)(*1*)(*3*)(*5*)] and 8.

12. Even with the model settings being the same throughout the 200-cyce run, the energy-content curve displays *variability* about the mean. The overall pattern of the curve suggests that the planet's climate system (i.e., energy content) appears relatively stable. Assuming such a "*steady state*" condition was achieved, it can be expected that the rate at which energy is leaving the system to space would be [(*less than*)(*equal to*)(*more than*)] the rate of incoming energy from space.

Keeping other settings the same, switch to the Random mode. Try several runs of the model to see differences and similarities in results. Since the settings were kept the same, the differences you observe, that is, differences in the means and departures from the means, must be due exclusively to chance within the model's operation. These can be referred to as examples of *natural variability* as they cannot be attributed to any change in the system settings (because there were no changes). That is, they were due to the inherent randomness built into the rules on which the model is based.

We will return to the AMS CEM in future investigations to follow the flow of energy through Earth's climate system in different simulations under different sets of conditions. We will then be observing evidence of *climate change*.

Earth's Climate System Models

The purpose of the AMS CEM is to provide a tool enabling you to explore fundamental aspects concerning energy flow to, through, and from Earth's climate system. Climate system models for scientific research and prediction are much more complex. They are mathematical computer-based expressions of the conversions between heat and other forms of energy, fluid motions, chemical reactions, and radiant energy transfer.

The use of models to predict weather and investigate the Earth system and its climate system was one of the most immediate results of the invention of the computer and rapid development of computer technology beginning in the 1950s. NOAA's Geophysical Fluid Dynamics Laboratory (GFDL) at Princeton University created during the 1960s and 1970s what was generally recognized as the first true Global Circulation Model (GCM) that represented large scale atmospheric flow. It was at GFDL that the first climate change carbon-dioxide doubling experiments with GCMs were conducted.

Figure 3 schematically depicts the components, or sub-systems, of ***Earth's climate system*** (atmosphere, ocean, terrestrial and marine biospheres, cryosphere, and land surface) that must be considered in advanced computer climate models. These major components interact with each other through flows of energy in various forms, exchanges of water, the transfer of greenhouse gases (e.g., carbon dioxide, methane), and the cycling of nutrients. Solar energy is the originating source of the driving force for the motion of the atmosphere and ocean, heat transport, cycling of water, and biological activity.

13. The arrows in the figure identify the processes and interactions with and between the major components of Earth's climate system. The double-headed arrows show that [(***almost all***)(***about half***)(***few***)] of the processes and interactions between climate system components (e.g., precipitation-evaporation, land-atmosphere) involve bi-directional (upward/downward) flows.

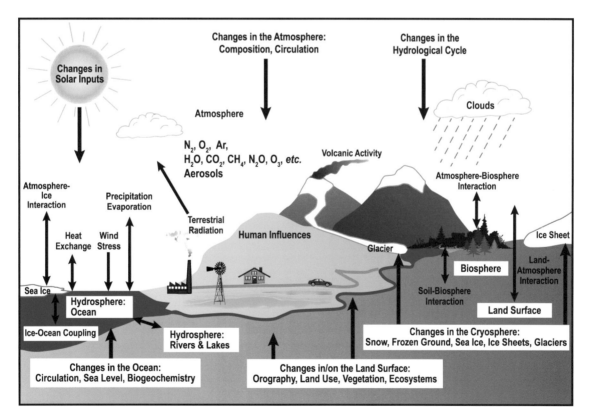

Figure 3.
Schematic view of the components of Earth's climate system, their processes and interactions. [IPCC AR4 WG1 faq-1-2-fig-1]

14. Six of the interactions depicted in Figure 2 are specifically labeled *"Changes in ..."* "Changes" imply forcing that results in climate change. While the human impact that most impacts global climate is change in the atmosphere (composition), the human impact most observably altering the local or regional climate is the one concerning the [(*ocean*)(*hydrological cycle*)(*cryosphere*)(*land surface*)].

Summary: This Investigation has presented the AMS CEM, a simple conceptual model that demonstrates climate as a planet system's response to external forcing (radiant energy from the Sun) and the amount of energy that is held in the system. It embodies some basic elements of computer-based climate models which are representations of the climate system based on the mathematical equations governing the behavior of the various components of the system, including treatments of key physical processes, interactions, and feedback phenomena.

EdGCM Project:

Computer-driven global climate models (GCMs) are prime tools used in climate research. The Educational Global Climate Modeling Project provides a research-grade GCM, called **EdGCM**, with a user-friendly interface that can be run on a desktop computer. Educators and students can employ EdGCM to explore the subject of climate change the way research scientists do. The model at the core of the EdGCM is based on NASA's Goddard Institute for Space Studies GCMs. To learn more about EdGCM, go to: *http://edgcm.columbia.edu/*.

Please note that the Internet addresses appearing in this Investigations Manual can be accessed via the "Learning Files" section of the course website. Click on "Investigations Manual Web Addresses." Then, go to the appropriate investigation and click on the address link. We recommend this approach for its convenience. It also enables AMS to update any website addresses that were changed after this Investigations Manual was prepared.

2A - 1

CLIMATE SCIENCE FROM AN EMPIRICAL PERSPECTIVE

Driving Question: *What observational weather data are collected to describe climate, and how are they organized and analyzed to provide an empirical view of a location's climate?*

Educational Outcomes: To identify the kinds of atmospheric data accumulated and analyzed to describe a location's climate. To learn where such information can be found for U.S. locations (including your own). To demonstrate how local climatic data can be analyzed to look for associations among weather elements.

This investigation examines the kinds of climatic data routinely acquired and analyzed by NOAA as the basis for describing climate at hundreds of locations nationwide. After completing this investigation, you should be able to:

- Describe and interpret information appearing in *Local Climatic Data, Annual Summary with Comparative Data* based on data collected at a local National Weather Service office.
- Explain how to access climate data from the National Climatic Data Center (NCDC).

Local Climatic Data

Climate data are systematically collected because they are extremely useful for many purposes. Farmers use their knowledge of weather and climate over a long period to determine what crops to plant and for guidance on when to plant and when to harvest. Utilities use climate data for planning production and distribution of energy supplies and the reallocation among types of such supplies. The building industry uses climate data in the design of structures, including their necessary strength, heating and cooling energy requirements, and the associated building codes that regulate them. Scientists modeling Earth's climate system employ derived products built on climatological data sets in model development and for verifying climate and climate change models. These are just a few of the uses of climate data.

In the U.S., weather data are gathered by NOAA's National Weather Service offices and other organizations and compiled at state, regional, and national centers for distribution to users. NOAA's National Climatic Data Center (NCDC) in Asheville, NC, is responsible for compiling and archiving U.S. data as well as being a depository for worldwide data on weather and the environment. This information, in turn, is made available to users on a variety of media.

A primary publication of NOAA's National Climatic Data Center (NCDC) based on weather data from local National Weather Service (NWS) offices is the *Local Climatological Data (LCD)*. It is published for about 275 NWS observing sites in monthly and annual summaries and available free online.

The *LCD, Annual Summary with Comparative Data*, for Grand Island, Nebraska (KGRI for the Year 2011 is used in this investigation). To retrieve Grand Island's LCD, go to: *http://www7.ncdc.noaa.gov/IPS/lcd/lcd.html*. Or, scan the QR code at the right.

QR Code

Select in sequence, "Nebraska", "GRAND ISLAND", and "2011-ANNUAL". Then, midway down the page, click on the URL provided.

[Alternate paths to the Grand Island 2011 LCD Annual Summary in the event the above does not provide access: *http://cdo.ncdc.noaa.gov/annual/2011/2011GRI.pdf* , or, *http://cdo.ncdc. noaa.gov/annual/2011/* and click on "2011GRI.pdf"]

Upon retrieval, you will be viewing the front page of the 2011 LCD Annual Summary for Grand Island. Either print out pages 1, 2, 3 and 7 of the document for reference or refer to them onscreen to respond to the following.

1. Examine the temperature graph appearing on the report's front page. The first of each month is indicated by the vertical dashed line directly above the monthly label, starting with 1 January at the left. Daily temperature ranges are plotted as vertical lines on the graph. The top end of each line signifies the maximum daily temperature and the bottom end reports the day's minimum temperature. The lowest minimum temperature for 2011 was –12 °F. It occurred on [(*12 January*)(*23 January*)(*6 December*)].

2. Locate on the temperature graph the highest maximum temperature reported at Grand Island for the Year 2011. It and the lowest minimum temperature for the year indicate that the annual temperature range at Grand Island in 2011 was about [(*85*)(*105*)(*115*)] F degrees.

3. A horizontal dark blue line is drawn on the temperature graph at 32 °F. Assuming that frost occurs if the temperature falls to 32 °F or lower, the approximate date of the last spring frost in 2011 at Grand Island when the minimum temperature dropped to 29 °F was about [(*10*)(*20*)(*28*)] April. (Note: There were four dates in May 2011 when the minimum temperature dropped to 33 °F.)

4. The green curve drawn across the Daily Max/Min Temperature graph presents the normal (or average) maximum temperatures for every day of the year. The brown curve delineates the normal daily minimum temperatures. These normal values are based on the average of maximum or minimum temperatures over a recent 30-year period (1971-2000). According to the graph, the date of the last spring frost in 2011 was [(*earlier than*)(*within a day or two of*)(*later than*)] normal.

5. The middle graph on the *LCD* front page reports daily precipitation in liquid equivalence. The maximum precipitation in one calendar day in 2011 occurred on about 20 May. According to the graph, the amount was approximately [(*1.3*)(*1.9*)(*2.4*)] inches.

Examine page 2 of Grand Island *LCD, Annual Summary*, entitled "METEOROLOGICAL DATA FOR 2011", and complete the following:

6. The lowest monthly average daily temperature ("Average Dry Bulb") for 2011 occurred during [(*January*)(*February*)(*December*)].

7. The average daily temperature during that month was [(*18.6*)(*22.0*)(*22.6*)] °F.

8. The highest monthly daily average temperature for the year (79.4 °F) occurred during [(*June*)(*July*)(*August*)].

9. The monthly precipitation (water equivalent) ranged from November's 0.21 in. to May's 8.70 in. The total 2011 precipitation was [(*8.70*)(*17.61*)(*27.16*)] in.

Examine page 3 of the *LCD, Annual Summary* entitled "NORMALS, MEANS, AND EXTREMES". The "Normals" presented are averages of observations taken during the 1971-2000 time period. "Mean" values are averages for the entire period of record of the weather element. **The values appearing on this page are commonly considered to be the "climate" of the station or region.** Complete the following:

10. The Highest Daily Maximum temperature ever recorded at Grand Island over its entire 65 years of record was [(*104*)(*109*)(*110*)] °F in August 1983.

11. The normal total annual precipitation at Grand Island is [(*13.96*)(*25.89*)(*36.27*)] in.

12. Comparing the actual annual value of a climate element with its normal value is a measure of *climate variability*. Comparison of the actual 2011 annual precipitation at Grand Island from Item 9 with the normal value in Item 11 shows a variability (or departure from the normal) of [(*1.27 in. below*)(*1.27 in. above*)(*0 in. from*)] normal.

13. Also included as part of the **LCD, Annual Summary** on page 7 is a brief narrative describing the location and climatic aspects of the area surrounding the local NWS office. Its climate is described as predominantly [(*maritime*)(*continental*)] in nature. According to the description, Grand Island is within 50 miles of the center of the coterminous U.S.

14. According to the narrative, incursions of maritime tropical air from the Gulf of Mexico [(*do*)(*do not*)] make it to Grand Island.

15. The Grand Island narrative also describes evidence of local climate change due to anthropogenic activities, including increased farm irrigation and use of soil management techniques. The detected climate change includes [(*reduced dust storms*)(*higher growing season humidities*)(*both of these*)].

Acquiring LCDs

The NCDC website you have already visited (*http://www7.ncdc.noaa.gov/IPS/lcd/lcd.html*) is your portal to all **LCD, Annual Summary with Comparative Data** and **LCD, Monthly Summary** publications.

From the NCDC website, examine both the **Annual Summary** and a **Monthly Summary** for the location closest to you. Examine its contents as you expand your knowledge of your local climate.

You also can acquire local climate data based on observations made as recently as yesterday. Go to your local NWS office website via *http://www.nws.noaa.gov/organization.php*. At your local station's website, under the Climate heading to the left, click on "Local". Select the product, location, and timeframe for the data you are seeking. Then click on Go.

Climate Normals Update

Climate Normals are three-decade averages of climatological variables, including temperature and precipitation, used as references for comparing observational data. These are updated at the end of each decade. NOAA's National Climatic Data Center (NCDC) released the 1981-2010 Normals on July 1, 2011, replacing the 1971-2000 Normals used through 2011. Comparison of selected 1971-2000 Annual Normals with 1981-2010 Annual Normals for Grand Island, Nebraska follows in **Table 1**:

Table 1. Comparison of selected "Old" and "New" Grand Island Annual Normals

Element	Annual Normals 1971-2000	Annual Normals 1981-2010
Maximum Temperature (°F)	61.1	62.4
Mean Temperature (°F)	49.9	50.8
Minimum Temperature (°F)	38.6	39.3
Cooling Degree Days	1027	1035
Heating Degree Days	6385	6198
Precipitation (in.)	25.89	26.66
Snow (in.)	32.9	29.0

16. Comparison of the 1971-2000 and 1981-2010 Annual Normals shows that Grand Island's climate normals [(*remained the same*)(*changed*)] from the earlier period to the later period.

17. The changes, if any, in temperature or temperature-related annual normals at Grand Island are consistent with that of a [(*cooling*)(*steady*)(*warming*)] local climate.

Figure 1 provides a comprehensive color-coded view of temperature changes between the 1971-2000 and 1981-2010 Normals.

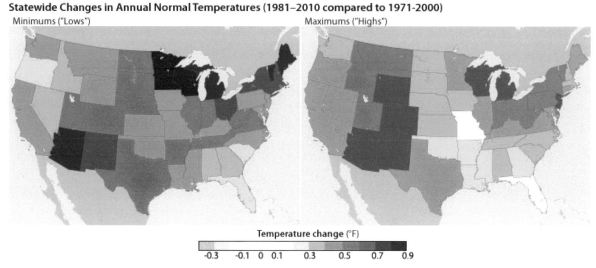

Figure 1.
Changes in annual normal temperatures. [NCDC/NOAA]

18. Comparison of Grand Island's "old" and "new" temperature normals given in Table 1 with Figure 1 shows that Grand Island's local change trends [(*were*)(*were not*)] generally consistent with state, regional (several-state), and coterminous U.S. changes.

19. Figure 1 reveals a pattern of changes in annual normal minimum and maximum temperatures at state levels across the coterminous U.S. that is best described as general [(*cooling*)(*mixed cooling and warming*)(*warming*)].

20. Figure 1 shows that the greatest temperature changes occurred in the statewide annual normal [(*minimum*)(*maximum*)] temperatures.

Summary:

Observational data are the feedstock of the processes employed to describe climate from an empirical perspective. Included in the analysis of such data are the calculations of normals, typically covering the most recent three decades, to provide a frame of reference for evaluating individual observations and for looking for trends. While the focus in this investigation has been on U.S. climatic data, countries worldwide are also engaged in climatic data gathering under the auspices of the World Meteorological Organization.

A word of caution: What might be happening at the local scale should not simply be extrapolated to the state, regional or global scale, or vice versa. Climate change, if any, is not uniform across local, regional, and global scales. Local climate warming or cooling might be the reverse of what is happening at other localities or at regional or global scales.

Investigation 2B: CLIMATE VARIABILITY AND CHANGE

2B - 1

Driving Question: *What is climate change? Is there short-term evidence that human activity can modify climate? How can we objectively determine modification of climate due to human activity?*

Educational Outcomes: To describe what is meant by climate variability and climate change. To describe how human activities can significantly change one or more climate measures, and how stopping particular human impacts might result in the climate measures returning to their original states.

Objectives: After completing this investigation, you should be able to:

- Distinguish between climate variability and climate change.
- Describe one instance of climate change likely to be caused by human activity.

Materials: Red and blue pencils, ruler or other straight edge.

Climate Variability and Climate Change

Earth's climate changes when the amount of energy contained in its climate system varies. Determinations of whether or not Earth's climate has changed, is changing, or is likely to change are elusive tasks. While recognizing that the geological and historical record shows an evolving climate, we face daunting challenges in our attempts to evaluate recent climate trends (e.g., global temperature rise) and computer climate model products as evidence of short-term variations in the climate or of persistent change in the climate system. **When we study climate, we must ask, "Are we observing statistical fluctuations of climate measures in a steady climate state or are we witnessing real change in the mean climate state?"** We can start looking for answers to this question by defining what we mean by climate variability and climate change. The more precisely we describe what we are looking for, the more likely we will know when we find it.

Climate variability refers to variations about the mean state and other statistics (such as standard deviation, statistics of extremes, etc.) of the climate on all time and space scales beyond that of individual weather events [adapted from IPCC]. It is often used to describe deviations in climate statistics over a period of time (e.g., month, season, year) compared to the long-term climate statistics for the same time period. For example, a particular year's average temperature will very likely differ from the mean annual temperature for a recent 30-year period. Such variability may be due to natural internal processes within the climate system or to variations in natural or anthropogenic external forcing.

Climate change refers to any change in climate over time, whether due to natural forcing or as a result of human activity [adapted from IPCC]. It refers to a significant change in the climatic state as evidenced by the modification of the mean value or variability of one or more weather measures persisting over several decades or longer. **Global climate change occurs ultimately because of alterations in the planetary-scale energy balance between incoming solar energy and outgoing heat energy (in the form of infrared radiation).** The mechanisms that shift the global energy balance result from a combination of changes in the incoming solar radiation, changes in the amount of solar radiation scattered by the Earth system back to space, and adjustments in the flow of infrared radiation from the Earth system to space, as well as by changes in the climate system's internal dynamics. Such mechanisms are termed *climate forcing mechanisms*.

Climate change can occur on global, regional, and local scales. The prime (and most pressing) example of climate change is global warming, recognized almost universally as mostly due to increasing atmospheric carbon dioxide via burning of fossil fuels. Among examples of anthropogenic climate forcing at a more regional level include changes in Earth's surface reflection of sunlight back to space due to land use and even the subtle impact of aircraft contrails (as will be examined later in this investigation). Local climate data can provide evidence of climate modification through human activity as seemingly innocuous as changes in local farming practices. As you recall, the narrative section of the National Climatic Data Center's *Local Climatic Data (LCD), Annual Summary* for Grand Island, Nebraska, reports that the increased use of irrigation and soil management techniques in local farming reduced the frequency of dry season dust storms while increasing growing-season atmospheric humidities.

Determining Climate Variability and Climate Change

The scientific, objective investigation of climate and climate change requires the use of clearly defined terms. We have already defined climate in terms of its empirical and dynamic aspects.

We will use the term *climate variability* to describe the variations of the climate system around a mean state (e.g., average temperature of a single month compared to the average monthly temperature for that month as determined from several decades of observations). Typically, the term is used when examining departures from a mean state determined by time scales from several decades to millennia or longer.

The AMS CEM can be employed to illustrate climate variability. Go to the course website and click on "AMS Conceptual Energy Model". Then click *Run the AMS CEM*. Set the model for one atmosphere, 100% Sun's energy, 200 cycles, and Introductory mode, and click on "Run". The on-screen visualization includes, above the graph, the Mean and Standard Deviation of energy units residing in the imaginary planet's climate system (surface and atmosphere) over the 200-cycle run. The graph displays curves drawn to report numbers of energy units in the planetary system at the end of each cycle (jagged blue curve) as well as the 5-cycle running mean (smoother green curve).

1. According to the CEM depicting the planet's climate system and space above, at the end of the 200th cycle there were [(*4*)(*5*)(*6*)] energy units residing in the climate system. Note that this is the same value for the 200th cycle as depicted in the graph below the window.

Figure 1 is an abridged version of the graph portion of the on-screen image that displays the jagged blue curve reporting the number of energy units in the planetary climate system at the end of each cycle.

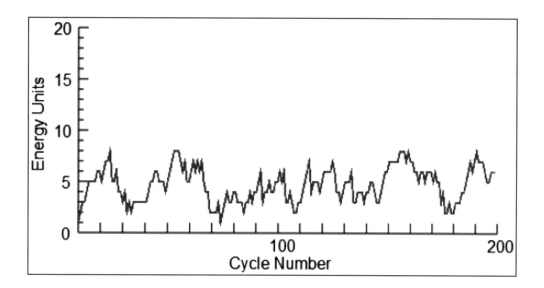

Figure 1.
Energy units residing in Earth system over 200-cycle run of AMS CEM.

2. According to Figure 1, and ignoring the initial "spin up" of the model, the number of energy units residing in the planetary climate system at the end of each cycle rose and fell between [(*1 and 7*)(*1 and 8*)(*2 and 9*)]. The range (the largest minus the smallest in a set of values) was 7. Range is a measure of variability.

3. In the on-screen image, note the mean number of energy units in the planetary climate system for the 200 cycles as reported above the graph. On Figure 1, draw a solid horizontal straight line representing that mean value of [(*1.66*)(*3.2*)(*4.7*)] energy units.

4. Shade with colored pencils the areas between the line depicting the mean and the jagged blue curve. Color those areas above the mean red and those below the mean blue. Visually, it should be apparent that the total shaded area above the mean line is equal to the total shaded area below the mean line. The departures of the jagged curve from the mean line represent the energy-unit variability of the system for that particular 200-cycle run. A statistical measure of the magnitude of this variability is **standard deviation** (SD). The greater the spread of observed values from the mean, the greater the SD. According to the on-screen image, this 200-cycle run exhibits a SD of [(*1.66*)(*3.2*)(*4.7*)].

5. Draw on the Figure 1 graph horizontal dashed lines representing +1 SD (mean plus SD value = 6.36) and –1 SD (mean minus SD value = 3.04) from the mean. **Figure 2** shows a plot of a normal distribution by SD (or σ). [A normal distribution is a frequency graph

of a set of values, usually represented by a bell-shaped curve symmetrical about the mean (μ).] According to the Figure 2 graph, [(*4.1%*)(*68.2%*)(*95.4%*)] of the observed values fall between +1 SD and –1SD of the mean. On Figure 1, compare the shaded areas between +1 SD and –1 SD with the total shaded area to confirm that this appears to be correct.

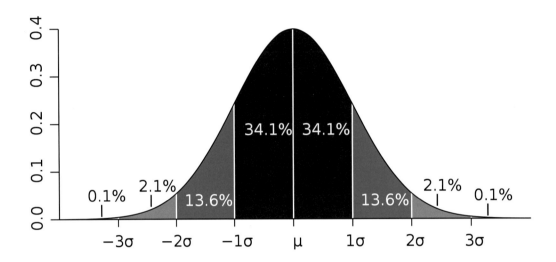

Figure 2.
Normal Distribution by Standard Deviation (SD or σ – Greek letter Sigma) [Mwtoews, Wikipedia]

6. Draw horizontal dashed lines on Figure 1 representing +2 SD (mean plus 2 SD = 8.02) and –2 SD (mean minus 2 SD = 1.38). Figure 2 shows that [(*4.1%*)(*68.2%*)(*95.4%*)] of observed values can be expected to fall between +2 SD and –2SD. On Figure 1, compare the shaded areas between +2 SD and –2 SD with the total shaded area to confirm that this appears to be correct.

7. Figure 2 presents percentage values within SD intervals that show 4.4% of observed values can be expected to have values greater than +2 SD or lower than –2 SD from the mean. Your analysis of the Figure 1 curve [(*does*)(*does not*)] show observed values beyond 2 SDs.

8. Because the CEM settings were kept the same throughout the 200-cycle run (one atmosphere, 100% Sun's energy), the variability observed [(*is*)(*is not*)] due to a change in the system. So, it can be described as natural variability.

Climate Change?

Climate change, as used in this course, refers to any sustained change in the long-term statistics of climate elements (such as temperature, precipitation or winds) lasting over several decades or more, whether due to natural variability or as a result of human activity. This definition follows the ***AMS Glossary of Meteorology, 2nd edition, 2000,*** and that used by the Intergovernmental Panel on Climate Change (IPCC). (While this course employs the definition given here, keep in mind that climate change is defined by some to mean a change of climate that can be attributed directly or indirectly to human activity only. The context in which the term appears will usually inform the reader of the definition employed.)

9. Determining whether or not climate change has occurred requires comparison of **[(*climate means*)(*climate variability*)(*both of these*)]** as determined from empirically acquired climatic data for the same locality.

Return to the AMS CEM. Set the model for two atmospheres, 100% Sun's energy, 200 cycles, and introductory mode, and click on "Run". Only the atmosphere setting is different from the AMS CEM scenario examined in the first part of this investigation when it was set at one atmosphere. The two-atmosphere setting can be thought of as representing a doubling of atmospheric CO_2 compared to the one-atmosphere setting.

10. Compare this model product with that of the one-atmosphere model run. With two atmospheres, the mean is **[(*2.36*)(*4.7*)(*10.7*)]** energy units. This higher mean, compared to the one-atmosphere model run, suggests the doubling of the atmosphere has brought about a sustained change in the planetary climate system, that is, it seems to exemplify climate change.

11. The two-atmosphere model run produced a SD of **[(*2.36*)(*4.7*)(*10.7*)]**.

Knowing the means, SDs, and numbers of cycles for the one- and two-atmosphere scenarios, a statistical test can be applied to make a determination to some level of confidence that the differences in the two model products are due to other than chance. The **Student's *t*-test** is commonly employed to make such a determination. Go to: *http://www.graphpad.com/quickcalcs/ttest1.cfm*. To use this *t*-test calculator, in Step 1 click on the "Enter mean, SD and N" button, in Step 2 enter under Group 1 one-atmosphere values of mean, SD, and N (200) used earlier in this investigation, and under Group 2 enter the two-atmosphere values of mean, SD, and N (200). After being sure "Unpaired *t* test" is selected in Step 3, click on "Calculate now" in Step 4.

12. The Unpaired *t* test results that appear report a two-tailed P value of less than **[(*0.0001*)(*0.001*)(*1.0*)]**. The P value is a probability, and can have a value ranging from zero to one. The smaller the P value, the greater the probability that the difference between sample means is due to something other than chance or coincidence. The smaller the P value, the more confident you can be that the two samples you compared are from different populations, that is, they are significantly different.

13. On the same Unpaired *t* test results page, the P value reported indicates the difference between the means of the two samples is considered to be statistically **[(*not significant*)(*significant*)(*extremely significant*)]**.

14. This indicates with a high level of confidence that the difference of the one-atmosphere and two-atmosphere scenarios is due to some factor, not simply chance or coincidence. We can therefore say or infer with a high level of confidence that the addition of a second atmosphere (or doubling of CO_2) resulted in [(*__no climate change__*)(*__climate change__*)].

Optional: To become more familiar with these basic statistics being applied to the AMS CEM output, compare different pairs of runs of the model including using the "random" mode and changing other settings (one at a time). Make the comparisons by calculating the t test.

Aircraft Contrails, Cirrus Clouds, and Climate Variability and Change:

The advent of jet aircraft and the huge growth in air traffic after World War II resulted in an increase in cirrus clouds formed by contrails from engine exhaust. A question of considerable interest to atmospheric scientists has been whether or not the increase in contrails and related cirrus clouds has impacted weather and climate. If the increase in contrails has impacted climate, it is obviously an example of anthropogenic climate change. **Figure 3** shows the prevalence of contrails as well as indications that under certain conditions contrails seed a more expansive cloud cover.

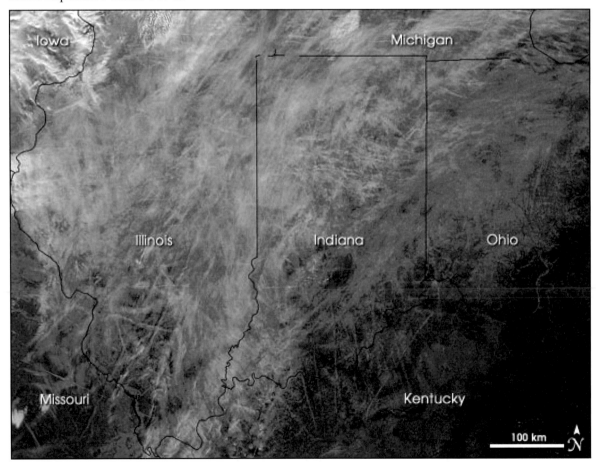

Figure 3.
NASA MODIS image of contrails over Midwestern U.S. [NASA] *http://earthobservatory.nasa.gov/IOTD/view.php?id=7161*

Aircraft contrails are clouds that form when hot jet engine exhaust containing considerable quantities of water vapor (a combustion product) mixes with the cold high-altitude air. They are the most visible anthropogenic atmospheric constituents in regions of the world with heavy air traffic.

Because contrails and related cirrus clouds triggered by contrails reflect solar radiation and absorb and emit infrared radiation, it is reasonable to expect that their persistence and pervasiveness likely impact climate. They not only may be affecting climate at the current time, but their impact can be expected to increase as it is projected that jet air traffic will grow in the decades ahead.

A Possible "9-11" Climate Lesson

The date 11 September 2001, or simply "9/11," marks one of the darkest days in our country's history. By 9:30 EDT on that morning, it became clear to officials at the Federal Aviation Administration (FAA) that something was terribly wrong. One immediate response to prevent the possibility of other aircraft being used as destructive weapons was the remarkably quick national grounding of all commercial, military, and private aircraft. Within an hour or so, the more than 4000 aircraft in U.S. airspace and international flights headed to this country were directed to the airports nearest them. By the afternoon of 9/11, the only contrails visible on satellite images were those coming from the President's Air Force One and its two fighter jet escorts on their way to Washington, DC.

U.S. skies remained essentially clear of aircraft for more than a day. In general, the grounding remained in effect until 13 September. The grounding totally ended when Washington's Reagan National Airport finally opened on 4 October.

The 9/11 aviation shutdown gave scientists unique opportunities to study a few isolated contrails developing without interference from neighboring contrails and to acquire evidence of possible climate shifts. A study of particular significance to climate change was conducted by Prof. David Travis (University of Wisconsin-Whitewater) comparing surface air temperatures across the country during the aircraft grounding with those before 9/11.

The Travis analysis showed that during the absence of contrails (the 11-13 September time period when skies were generally clear), the difference between the highest temperature during the day and the lowest temperature at night increased 3 Fahrenheit degrees on average and as much as 5 Fahrenheit degrees in areas of the country where contrails were usually most common. This led Travis to conclude that contrails and related cirrus clouds influenced climate by increasing the reflection of incoming solar radiation back to space during the day, thereby reducing heating at Earth's surface, and then absorbing some of the upwelling infrared radiation from Earth's surface at night. A considerable amount of the absorbed radiation is emitted back towards Earth's surface, where it has a heating effect. Together, these two processes potentially reduce the diurnal (daily) temperature range. Travis speculated that climatologically there is a net cooling effect because there are generally more flights and contrails during the day than at night.

15. Recalling the perspectives of climate as described in Investigation 1A, Prof. Travis's study of observed surface air temperatures across the country was essentially a(n) [(*dynamically*)(*empirically*)] based investigation.

16. Because of the distribution of contrails as displayed in Figure 3, any climate change brought on by the occurrence of contrails might best be described as [(*regional*)(*global*)] in scale.

While Travis's statistical treatment of climatic data in his study does show a greater temperature range and a higher mean temperature when contrails were absent, it is not clear that the evidence demonstrates unequivocally the impact of contrails on climate. A 2008 study, "Do contrails significantly reduce daily temperature range?" by Gang Hong et al, Texas A&M University, reports that the increase of the average daily temperature over the United States during the 11-14 September 2001 aircraft grounding period was within the range of natural temperature variability observed from 1971 to 2001.

Hong's study concluded that the missing contrails may have affected the daily temperature range, but their impact is probably too small to detect to a level of statistical significance. Hong showed that the diurnal temperature range is governed primarily by lower altitude clouds, winds, and humidity. Specifically, the unusually clear and dry air masses covering the Northeastern U.S. in the days following the terrorist attacks favored unusually large daily temperature ranges.

Summary: The studies referred to in this investigation are presented to demonstrate the challenges of identifying and discriminating between natural climate variability and climate change. The AMS CEM was employed to illustrate climate variability and climate change. Studies of the 11-14 September 2001 time period when contrails were temporarily absent over the U.S. potentially provided a unique opportunity for detecting climate change, if any, due to contrail impact. Careful studies of probable causes of unusually large daily temperature ranges at the time ascribe whatever differences that were detected as explainable within the range of natural variability of atmospheric conditions.

SOLAR ENERGY AND EARTH'S CLIMATE SYSTEM

Driving Question: *How does solar energy received by the Earth vary during the year at different latitudes?*

Educational Outcomes: To describe how the amount of solar radiation intercepted by Earth varies at different latitudes over the period of a year. To learn how changes in the Sun's path through local skies impact the amount of incident solar radiation. To make comparisons of how much solar radiation is received at tropical, midlatitude, and polar locations at different times of the year. To estimate the impact of the atmosphere on incoming solar radiation by comparing the amount received at Earth's surface with that striking the top of the atmosphere.

Objectives: As stated in *Investigation 1A*, climate can be thought of as the story of solar energy intercepted by Earth being absorbed, scattered, reflected, stored, transformed, put to work, and eventually emitted back to space as infrared radiation. The solar energy entering the Earth system is the ultimate boundary condition of climate as the Sun is the source of energy that heats Earth's climate system.

After completing this investigation, you should be able to:

- Describe the variation of solar radiation received at the top of the atmosphere at equatorial, midlatitude, and polar locations over the period of a year.
- Compare the amounts of solar radiation received at a midlatitude location at the top of the atmosphere and at Earth's surface under clear-sky and average conditions during different times of the year.

Incoming Solar Radiation

Over the period of a year the amount of solar radiation received at Earth's surface varies considerably at most latitudes. This variation is governed largely by the boundary condition arising from the changing paths of the Sun through the local sky, due to Earth rotating on an axis inclined to the plane of its annual orbit about the Sun. These planetary motions constantly change the part of Earth's surface bathed by the Sun's rays. Every latitude on the globe has its own annual pattern of incident sunlight. These patterns are governed by the shifting path of the Sun through the local sky and the lengths of daily periods of daylight. These variations are primarily responsible for the unequal distribution of absorbed solar energy from the equator to the poles that drives Earth's climate system.

Figure 1 displays the approximate paths of the Sun through the local sky at (A) the equator, (B) a middle latitude location in the Northern Hemisphere, and (C) the North Pole on the first days of summer, fall, winter, and spring. The paths are continuously changing through perpetual annual cycles. Among the variations by latitude, at the equator the periods of

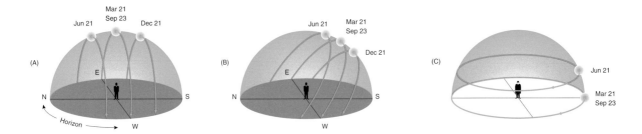

Figure 1.
Path of Sun through the sky on the solstices and equinoxes at (A) the equator, (B) a middle latitude location in the Northern Hemisphere, and (C) the North Pole.

daylight are a constant half-day throughout the year, while at the pole there is <u>one</u> period of sunlight through the year that lasts continuously for six months.

At the top of the atmosphere when Earth is at its mean distance from the Sun, an average of about two calories of solar energy per minute strikes a square centimeter (1.4 kilowatts per square meter or 0.14 watt per square centimeter) of a flat surface oriented perpendicular to the Sun's rays. This average rate is called the **solar constant**. The rates at which solar energy penetrating the atmosphere actually strikes Earth's surface are quite different and highly variable. At any instant, the only point at which the Sun's rays are perpendicular to Earth's surface is where the Sun is in the zenith. That sub-solar point races steadily around the planet once a day as it follows an annual path that spirals between 23.5 degrees North and 23.5 degrees South latitudes. Twice a year, on the vernal and autumnal equinoxes, the Sun is positioned directly above the equator. During the time between the vernal and the succeeding autumnal equinox the sub-solar point is located in the Northern Hemisphere, while from the autumnal to the following vernal equinox, the sub-solar point is positioned in the Southern Hemisphere. The solstices occur when the sub-solar point reaches its maximum latitude positions (23.5° N on the first day of the Northern Hemisphere's summer in late June and 23.5° S on the first day of our winter in late December).

The Earth, spinning on an axis inclined 23.5 degrees from a line perpendicular to the plane of its orbit, presents an ever changing face to the Sun. Wherever daylight occurs, the path of the Sun through the local sky changes from day to day. Except at the equator, or at high latitudes when there are days of continuous daylight or darkness, the daily length of daylight also changes. Clouds, air molecules, and aerosols (tiny particles suspended in air) reduce the amount of solar radiation reaching Earth's surface. Some solar radiation is absorbed by atmospheric components and some is scattered or reflected back to space.

Because absorbed solar radiation is the fundamental driver of Earth's climate system, the purpose of this activity is to investigate the variability of solar radiation received at the top of the atmosphere and at Earth's surface at different latitudes over the period of a year.

The great variation in the solar radiation received at different latitudes throughout a year is primarily responsible for the temperature contrasts that result in the fluid parts of the Earth system (atmosphere and ocean) transporting huge quantities of heat energy from lower to higher latitudes. These energy flows accompany the weather patterns and temperature variations that characterize the seasons.

Effects of Latitude on Incoming Solar Radiation

What impacts do latitude and the atmosphere have on incoming solar radiation? NASA provides monthly averaged top-of-atmosphere insolation values as well as those incident on a horizontal surface at Earth's surface for any global location (*http://eosweb.larc.nasa.gov/cgi-bin/sse/sizer.cgi?email=na*). The term *insolation* is short for *in*coming *sol*ar radi*ation*. The top-of-atmosphere values would be the amount received at Earth's surface if there were no atmosphere. The NASA data report incident solar radiation in units of kilowatt hours per square meter per day (kWh/m^2/day). [For conversion purposes, 1 cal/cm^2 = 0.01 kWh/m^2.]

Figure 1(A) shows daily paths of the Sun at an equatorial location on the solstices and equinoxes. **Figure 2** displays a red curve that depicts the daily average solar radiation striking a horizontal surface at the top of the atmosphere over a location on the equator (0° Latitude). Data of average daily values for each month are plotted at mid-month. <u>Draw and label straight vertical lines on Figure 2</u> on the approximate dates of the Northern Hemisphere's vernal equinox (21 March), summer solstice (21 June), autumnal equinox (23 September), and winter solstice (21 December).

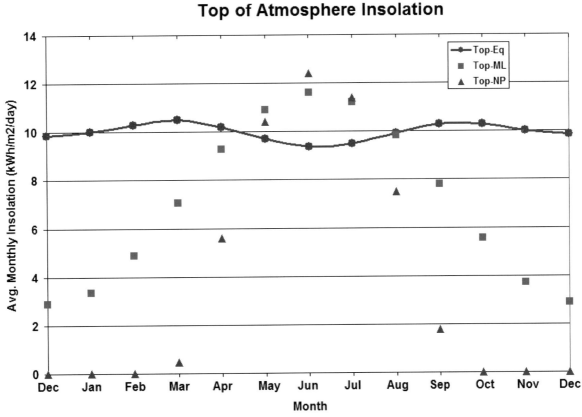

Figure 2.
NASA-generated average monthly top-of-atmosphere solar radiation (in kWh/m^2/day) at equator (Eq), midlatitude (ML), and North Pole (NP).

1. Note the two maxima and two minima portions of the Figure 2 equatorial insolation curve over the period of a year. Figure 2 shows that the minimum portions of average daily solar radiation values at the top of the atmosphere over the period of a year occur during the months of [(*__March and September/October__*)(*__June and December__*)]. These occur when the sub-solar point is at its highest latitudes for the year, as seen in Figure 1(A).

2. The equatorial top-of-atmosphere curve in Figure 2 shows that the two insolation maxima occur near the times of the [(*__solstices__*)(*__equinoxes__*)]. Because places on the equator experience essentially the same period of daylight (approximately 12 hours) every day of the year, the changes in the top-of-atmosphere radiation values over the year must be due primarily to changes in the path of the Sun through the local sky (that is, changes in the maximum daily altitude of the Sun). Note that the top-of-atmosphere insolation values do not vary greatly over the year at the equator, with the minimum insolation value being about 90% of the maximum insolation value.

3. **Figure 1(C)** describes Sun's path through the local sky at the North Pole (90°N) on the summer solstice and the equinoxes. It shows that on these days, the Sun is on or above the horizon for [(*__0__*)(*__12__*)(*__24__*)] hours. It can be inferred from the drawing that the Sun is below the local horizon continuously from the fall equinox to the next spring equinox (assuming no atmospheric effects).

4. Plotted on Figure 2 are data points (▲) of monthly average daily top-of-atmosphere insolation at the North Pole (90° N). Assuming top-of-atmosphere insolation is zero on the equinoxes, connect the adjacent values by drawing a smoothed curve (with a blue pencil if available). The curve shows the North Pole receives essentially no incoming sunlight for [(*__0__*)(*__3__*)(*__6__*)(*__9__*)(*__12__*)] months a year. (The insolation pattern at the South Pole is similar except six months out of phase with that at the North Pole.)

5. Plotted on Figure 2 are data points (■) of monthly average daily top-of-atmosphere insolation at 45 degrees North Latitude (45° N). Connect the adjacent values by drawing a smoothed curve (with a green pencil if available). The midlatitude curve makes it evident that the maximum insolation occurs on the summer solstice and minimum insolation occurs on the winter solstice. It can be determined from Figure 2 that on an average day in December the midlatitude location receives about [(*__25%__*)(*__50%__*)(*__75%__*)(*__100%__*)] of the top-of-atmosphere insolation it receives on an average June day.

6. The midlatitude curve is characterized by having one maximum value and one minimum value per year. Comparison of the curves now appearing on Figure 2 shows that the midlatitude location [(*__always__*)(*__sometimes__*)(*__never__*)] receives top-of-atmosphere insolation that is greater than what is received at the equator.

7. **Figure 1(B)** depicts Sun's paths through the midlatitude local sky on the solstices and equinoxes. The solstice paths demonstrate that changes in [(*__Sun's maximum altitude__*)(*__length of Sun's path__*)(*__both of these__*)] contribute to the midlatitude's annual range of daily insolation.

8. Comparisons of the three top-of-atmosphere curves drawn in Figure 2 demonstrate why latitude is considered as a fundamental control of climate. The insolation values over the period of a year vary the most at the [(*equatorial*)(*midlatitude*)(*polar*)] location.

9. The [(*equatorial*)(*midlatitude*)(*polar*)] location experiences the least change in insolation, over the period of a year.

10. The three top-of-atmosphere insolation curves in Figure 2 indicate that during the year, the most intense daily insolation occurs at the [(*equatorial*)(*midlatitude*)(*polar*)] location.

Impacts of the Atmosphere on Incoming Solar Radiation

The rate at which solar energy is received at the top of the atmosphere is the fundamental boundary condition of Earth's climate system. **Figure 3** is presented to demonstrate the impact of the atmosphere on the sunlight entering the Earth system at a 45°N location (Salem, Oregon). The red curve drawn on the figure represents the same top-of-atmosphere insolation data for 45° N as seen in Figure 2.

11. The blue data points (■) plotted on Figure 3 are calculated average monthly values of daily insolation received at Earth's surface at a 45° N location that would be observed under clear-sky conditions (no clouds). Create the annual clear-sky insolation curve by drawing a smooth curve through the data points (using a blue pencil if available). The figure shows that during the month of May, 10.9 kWh/m^2 of energy was received at the top of the atmosphere. The figure also shows that under continuous clear-sky conditions, it would receive about [(*4.1*)(*6.1*)(*8.1*)] kWh/m^2 on an average May day.

12. The May data shows that for that month about [(*26%*)(*41%*)(*76%*)] of the solar energy striking the top of the atmosphere is blocked by the clear atmosphere. This attenuation (loss) is due to solar radiation being absorbed by atmospheric gas molecules (H_2O and O_3) and particulates (dust) and by backscattering to space. Evaluation of the Figure 3 data for all 12 months produce about the same result. Global energy budget studies place the worldwide average for clear-air absorption and backscatter at about the same (actually, about 2% less).

13. The green data points (▲) plotted on Figure 3 are calculated average monthly values of daily insolation received at Earth's surface at a 45° N location that would be observed under average conditions (including clouds). Create the annual average surface insolation curve by drawing a smooth curve through the data points (using a green pencil if available). According to Figure 3, the May value for solar energy actually arriving at Earth's surface is [(*4.1*)(*5.1*)(*8.1*)] kWh/m^2 for an average day.

14. Compared with the May top-of-atmosphere value, it shows that the actual atmosphere (including clouds) blocks about [(*37%*)(*53%*)(*76%*)] of the solar energy entering Earth's atmosphere at Salem, OR.

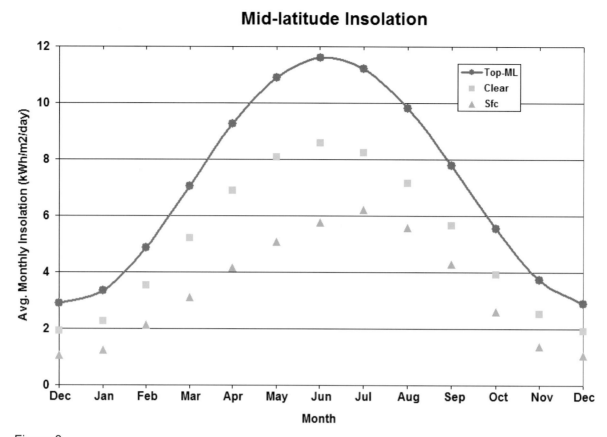

Figure 3.
NASA-generated daily averaged top-of-atmosphere, clear-sky and surface incident solar radiation data at Salem, OR, 45° N (in kWh/m²/day).

Global energy budget studies place the worldwide average for actual atmospheric absorption and backscatter conditions (including clouds) at near, but slightly less than, the Salem, OR, May value as reported in Figure 3. Figure 3 shows that the differences between Salem's May top-of-atmosphere and clear-sky values (2.8 kWh/m²/day) and the difference between clear-sky and actual surface insolation values (3.0 kWh/m²/day) are nearly equal. This infers that clear air and cloudiness are both major factors in determining how much of the solar radiation arriving at the top of Earth's atmosphere is attenuated before reaching the surface.

15. Actual surface insolation data reported in Figure 3 shows that monthly average values at Salem, OR range from 1.1 kWh/m²/day in December to [(*4.1*)(*5.1*)(*6.2*)] kWh/m²/day in July.

16. The actual surface insolation curve in Figure 3 demonstrates the wide swing in the amount of solar radiation arriving at Earth's surface at a midlatitude location over the period of a year. According to the data in the previous item, Earth's surface at Salem receives on an average July day about [(*2.1*)(*4.1*)(*5.6*)] times as much solar energy as on an average December day.

Summary: The story of climate starts with Earth's interception of solar radiation. The amount of sunlight intercepted varies considerably by latitude and time of year. The amount that reaches Earth's surface is determined by astronomical factors (e.g., solar constant, spherical Earth, rotation) and attenuation by atmospheric effects (e.g., cloudiness, absorption, scattering). Incident solar radiation at a particular location has climate implications resulting from the magnitude, path of the Sun through the daytime sky, and duration of daylight.

The solar energy entering the Earth system is the ultimate boundary condition of climate as the Sun is essentially the only source of energy that heats the Earth system – particularly its land and water surfaces and atmosphere. The amount of solar radiation received at Earth's surface varies considerably at most latitudes over the period of a year, setting the stage for annual climate swings. Although solar energy is essential in fueling the climate system, the Sun's energy intercepted by Earth is only the beginning of the story of climate.

Investigation 3B: ATMOSPHERIC CO_2, INFRARED RADIATION, AND CLIMATE CHANGE

Driving Question: *How does increasing the amount of carbon dioxide (CO_2) in the atmosphere impact the absorption of infrared radiation (IR) and average global surface temperature?*

Educational Outcomes: To describe the greenhouse effect of absorption and emission of infrared radiation (IR) by carbon dioxide (CO_2). To summarize from experimental evidence the changes in the absorption of IR as concentrations of atmospheric CO_2 increase. To demonstrate the impact of changing atmospheric CO_2 concentration on average global surface temperature as determined by a global climate model.

Objectives: After completing this investigation, you should be able to:

- Explain the greenhouse effect of IR absorption and emission by atmospheric CO_2 that results in higher atmospheric temperatures.
- Describe the impact of increasing the proportion of carbon dioxide in the atmosphere on absorption of IR and on temperature.
- Based on the investigation of CO_2, list fundamental understandings of how it and other greenhouse gases contribute to climate change.

The Atmospheric Greenhouse Effect

This investigation focuses on CO_2 to explore fundamental scientific understandings common to all atmospheric greenhouse gases.

Energy arrives at Earth as solar radiation, most intensely in the visible light portion of the electromagnetic spectrum. About half of the incoming radiant energy is reflected back to space (31%) or absorbed in the atmosphere (20%) while the other half is absorbed by Earth's surface. The heated surface then radiates IR upward, which is largely absorbed by atmospheric gas molecules including H_2O, CO_2, CH_4, and N_2O. These gas molecules emit the absorbed energy in all directions, half with an upward component and half in a downward direction, as demonstrated by the AMS CEM. By returning half of what they absorb back towards Earth's surface, these gases entrap heat within Earth's climate system. This is the ***greenhouse effect***.

Energy leaves the Earth system through the upper reaches of the atmosphere as IR. An increase in the concentration of greenhouse gases leads to temperatures in the troposphere (lower atmosphere) and at the surface increasing until a new planetary balance is achieved between solar radiation absorbed by the Earth system and IR exiting the Earth system.

1. As described above (and also a fundamental concept embodied in the AMS CEM), atmospheric greenhouse gas molecules absorb IR rising from Earth's surface and subsequently radiate IR [(***only downward***)(***randomly in all directions***)(***only upward***)].

2. Because of this radiation pattern, half of the IR emitted by the greenhouse gas molecules is directed earthward. The net effect is a trapping of heat energy in the Earth system and an increase in surface and tropospheric temperatures. If the proportion of greenhouse gases in the atmosphere increases but then steadies out to a constant value, the amount of IR escaping to space will adjust until a planetary balance exists between it and the amount of incoming solar radiation being absorbed by the Earth system. Under these conditions (and with a continuing steady supply of sunlight), global average surface and tropospheric temperatures will [(*increase*)(*remain the same*)(*decrease*)] over the long term.

3. An increase in the amount of a greenhouse gas in the atmosphere reduces the transmission of IR through the air at certain wavelengths. *Transmittance* is a measure of the percentage of electromagnetic radiation of a specified wavelength that passes through a substance, such as IR through the atmosphere. The less the transmittance, the greater the amount of IR that has been absorbed. This, in turn, causes a rise in global-scale tropospheric and surface temperatures. These temperature increases are associated with a(n) [(*increase*)(*decrease*)] in the transmittance of IR through the atmosphere.

Fundamentals of IR Absorption by Atmospheric Molecules

Atmospheric molecules are constantly vibrating, but they vibrate only in certain ways at certain frequencies depending on their unique compositions, masses and structures. These characteristics determine what wavelengths of electromagnetic radiation they will absorb and emit. The two most abundant atmospheric gases, nitrogen (N_2) and oxygen (O_2), exist as symmetric molecules which do not absorb or emit most radiant energy so that they are essentially invisible (or transparent) to the majority of both incoming sunlight and Earth's IR emissions. Their transmittance is essentially 100%. There are other gases, including CO_2, that are transparent to visible light but readily absorb and emit wavelengths of radiant energy in the IR portion of the electromagnetic spectrum. This is because they have vibrational modes that absorb energy in the IR wavelengths which Earth's surface radiates towards space. Increasing concentrations of these gases reduce the transmittance of IR at these radiation-sensitive wavelengths and, at high enough concentrations, can reduce transmittance to 0%. Their sensitivity at those IR wavelengths at which Earth's surface radiation is most intense sets the stage for the greenhouse effect.

To see the impact of CO_2 on the transmission of IR, go to: *http://chemistry.beloit.edu/Warming/pages/infrared.html*. Click on the graph to the upper right. The graph that appears shows data acquired with an IR spectrometer. A spectrometer is an instrument used to measure intensities of radiation over a specific portion of the electromagnetic spectrum. Absorption spectroscopy is employed by chemists and other scientists to identify and study chemicals, including gases. In this example, the IR spectrum is investigated in terms of how much of a pulsed IR signal is transmitted through a 10-cm path length across a range of wavelengths when the CO_2 concentration in a cell is varied.

4. In the graph, information along the x-axis is in terms of *wavenumber*, the number of wavelengths per cm. The larger the wavenumber, the shorter the wavelength along

the IR spectrum. As the wavenumber values decrease from left to right in the graph, wavelengths become [(*shorter*)(*longer*)] from left to right.

5. In the graph, the vertical scale plots transmittance of IR at different wavenumbers. Transmittance of 100% indicates total transparency, while transmittance of 0% indicates total absorption of the IR at that wavenumber. In the graph, at the lower right, is a window indicating the concentration of CO_2 in the infrared cell as a gas pressure in units of millimeters of mercury (mm Hg). The initial graph shows that when the CO_2 concentration is 0, the IR spectrometer indicates a transmittance of essentially [(*0%*)(*50%*)(*100%*)] across the range of IR wavenumbers covered by the graph.

6. To investigate transmittance as the concentration of CO_2 is changed in the IR cell, click on the ▶ button in the control bar below the graph. Each additional click on the button will present an IR spectrum for a greater concentration of CO_2. The graph shows that adding CO_2 causes transmittance of the IR signal through to the cell to be reduced [(*the same across all wavenumbers*)(*selectively at different wavenumbers*)].

7. At wavenumbers where there is a reduction in transmittance, it means there is an increase in the amount of IR absorbed. IR absorption speeds up molecular motions, which produces [(*a decrease*)(*no change*)(*an increase*)] in temperature in the cell. What has been observed in these laboratory measurements also happens in Earth's atmosphere. That is because any medium that absorbs IR is heated.

8. On the graph, step through increasing CO_2 concentrations while noting changes in transmittance at IR-sensitive wavenumbers. A 0% transmittance first occurs at a wavenumber of approximately [(*650*)(*2350*)(*3700*)] per cm. This is near the middle of a band of IR wavelengths most intensely emitted by Earth's surface.

9. Once a particular wavenumber or band of wavenumbers reaches a condition in which its transmittance has dropped to 0%, it is absorbing its maximum of IR. This condition is called **saturation**, and no more IR can be absorbed at the particular wavenumber or band of wavenumbers. Additional increases in CO_2 concentration in the IR spectrometer cell [(*will*)(*will not*)] result in additional absorption of IR at the saturated wavenumber(s).

10. On the graph, increase the CO_2 concentration over several steps while observing transmittance at the wavenumber where the transmittance curve first dropped to 0%. As the CO_2 concentration continues to increase after initial saturation, the width of the saturated wavenumber band along the base of the graph [(*decreases*)(*increases*)(*shows no change*)].

11. On the graph, examine the change in transmittance in the absorption band centering on the same wavenumber where the transmittance curve first dropped to 0% as the CO_2 increases from 0 to 804 mm Hg. It can be seen that as the CO_2 concentration increases, the width of the vertical band (bandwidth) of wavenumbers showing decreasing transmittance [(*broaden*)(*narrows*)], providing evidence of IR absorption. (Note also another intense absorption band near wavenumber 650).

The changes in these bandwidths partially explain why a greenhouse gas does not suddenly stop changing climate when it reaches concentrations that produce wavenumber saturation in the atmosphere. The boundaries of an absorption band are not abrupt. Though the core of an absorption band may be saturated, its edges are not. A wavenumber at which IR is absorbed is not immediately adjacent to a wavenumber that absorbs no IR. The IR-sensitivity of neighboring wavenumbers changes over a range of wavenumbers, grading gradually from maximum absorption to no absorption. Consequently, as CO_2 concentration increases, wavenumbers next to saturated wavenumbers absorb IR and the absorption bandwidth increases.

The increase of CO_2 concentration in an unsaturated environment (where transmittance values are high) has a substantially greater impact on IR absorption than at CO_2 concentrations at which some wavenumbers have become saturated (transmittance has dropped to 0% at those wavenumbers). The greatest impact of greenhouse gas molecules, including CO_2, on absorbing IR occurs at their lowest concentrations.

Laboratory-based absorption spectroscopy studies conclusively prove that CO_2 selectively absorbs IR. They also demonstrate that the efficiency of CO_2 in absorbing IR changes as the concentration of the gas varies; the greater the concentration, the lower the efficiency. Attendant with the absorption of IR is increasing molecular vibration and accompanying heating and temperature rise.

Impact of Increasing CO_2 Concentrations in the Atmosphere

The behavior of atmospheric CO_2 follows the same fundamental laws of physics and chemistry as does CO_2 studied in a laboratory setting. What is learned in the controlled laboratory environment makes it possible to conduct informed studies of the complex Earth-atmosphere system with its highly variable and complex flows of energy and mass.

The impact of increasing atmospheric CO_2 concentrations on the average global surface temperature can be explored with a climate model application from the University of Chicago based on the National Center for Atmospheric Research's Community Atmosphere Model (CAM) (*http://www.ccsm.ucar.edu/models/atm-cam*).

While you are invited to visit and manipulate the University of Chicago model yourself (*http://geoflop.uchicago.edu/forecast/docs/Projects/full_spectrum.html*), the data in **Table 1** were acquired from the model which produces equilibrium near-surface air temperature values when top-of-atmosphere radiation balance has been achieved. [For reference, the current CO_2 concentration is about 390 ppmv (ppmv = parts per million by volume). A model run with this value produces a 16.1 °C surface temperature.]

12. Table 1 data show that increasing the atmospheric CO_2 concentration results in a temperature increase. Comparison of the temperature change per 10 ppmv CO_2 increase between 0 and 80 ppmv shows that as the CO_2 concentration increases, the temperature change per 10 ppmv increment [(***increases***)(***decreases***)(***remains the same***)].

Table 1. Model output resulting from changing atmospheric CO_2 concentration.

CO_2 Concentration (ppmv)	Air Temperature (°C)	CO_2 Concentration (ppmv)	Air Temperature (°C)
0	−2.0	60	10.3
10	5.7	70	10.8
20	7.4	80	11.1
30	8.4	160	13.2
40	9.2	320	15.4
50	9.8	640	17.9

13. To determine the impact of doubling the proportion of CO_2 in the atmosphere, compare temperature changes calculated by the model. Note in Table 1 that when the CO_2 concentration doubles from 80 ppmv to 160 ppmv, the near-surface air temperature changes from 11.1 °C to 13.2 °C, an increase of 2.1 C degrees. Doubling the concentration again (from 160 to 320 ppmv), the model calculates a temperature change of [(*0.5*)(*2.2*)(*15.4*)] C degrees.

14. Assume this change in temperature with a doubling of CO_2 is representative of other model runs that double CO_2 concentrations. With this assumption, we can anticipate that if Earth's current atmosphere CO_2 concentration of 390 ppmv were doubled to 780 ppmv, Earth's near-surface temperature would increase at least [(*0.5*)(*2*)(*15*)] C degrees.

Summary

Some gases in the atmosphere, including H_2O, CO_2, CH_4, and N_2O, absorb IR rising from Earth's surface in a variety of wavenumber bandwidths unique to each gas. The absorbed energy is subsequently emitted in random directions with half directed downward and half upward. This trapping of heat energy in Earth's climate system is the greenhouse effect.

The amount of carbon dioxide in the atmosphere is increasing and its impacts as a greenhouse gas are increasing (although less efficiently). Comparison of IR spectra of CO_2 at different concentrations in the laboratory reveals the relative sensitivities of different wavenumbers to IR absorption and the broadening of absorption wavenumber bandwidths as concentrations increase. Increasing the CO_2 concentration means increasing absorption of IR and resulting temperature increases.

Changing the proportion of atmospheric CO_2 in a global climate model reveals a generalization concerning the effect of doubling the amount of CO_2 on temperature change. Doubling the CO_2, whether it is from 10 to 20 ppmv, 200 to 400 ppmv, or another doubling, produces about the same temperature change. That is, the lower the residual concentration of CO_2 in the atmosphere, the greater the greenhouse effect of a concentration increase.

This investigation centered on CO_2 as a greenhouse gas. However, other greenhouse gases exhibit the same general attributes. This is why gases such as CH_4 and N_2O, which occur in far smaller concentrations in the atmosphere, are much more efficient absorbers of IR and explains partially why H_2O, at much higher concentrations, is a less efficient absorber of IR.

Acknowledgement:

The carbon dioxide spectroscopy animation video was created by Prof. George Lisensky, Chemistry Department, Beloit College, Beloit, WI, from actual infrared spectra for those CO_2 concentrations. Used by permission: Copyright ChemConnections (*http://chemistry.beloit.edu*).

Investigation 4A: WATER, HEAT, AND HEAT TRANSFER

Driving Question: *How does water react to the gain or loss of heat?*

Educational Outcomes: To describe how different substances respond to the gain or loss of heat energy. Knowing the amounts and specific heats of different substances, determine their temperature changes as the result of adding or removing known quantities of heat energy. To explain latent heat, the heat energy that is added to or lost from a substance without a change in temperature during phase change (e.g., melting or freezing of water). To apply this knowledge about sensible heat and latent heat to determine how much heat energy is absorbed or released as the water substance changes temperature and phases.

Objectives: Water plays a central role in Earth's climate system. Water can exist as liquid, vapor, and solid within the temperature and pressure ranges existing at and near Earth's surface. Water is unique compared to most other substances in the relatively large gains or losses of heat energy required to undergo equivalent temperature changes. To change phase, water absorbs or releases huge amounts of energy compared to practically all other substances. These properties make water a primary vehicle for transferring heat energy from place to place in the Earth system. The resulting mass and energy flows constitute the global water cycle; a major component of Earth's climate system.

After completing this investigation, you should be able to:

- Describe temperature changes resulting from heat transfer to different substances.
- Describe the role of heat energy in the phase changes of water.
- Determine how much heat is involved in temperature and phase changes of water substance.

Sensible Heating and Temperature Change

Heat gains or losses cause substances to warm or cool (unless they are undergoing phase changes). Sensible heating is the heat transfer that causes a temperature change of a substance without change of phase. Sensible heating involves a property of a substance called its ***specific heat***, defined as the amount of heat required to change the temperature of 1 gram of a substance 1 Celsius degree. Listed in **Table 1** are specific heats of several substances in calories per gram per Celsius degree (**cal/g/C°**). [1 cal/g/C° = 4.186 J/g/C°]

Table 1. Specific heats (c) of some common substances (cal/g/C°)

solids		liquids		gases	
ice	0.5	**water**	1.0	**water vapor**	0.5
copper	0.1	alcohol	0.6	nitrogen	0.25
marble	0.2	mercury	0.33	carbon dioxide	0.2
sand	0.2				

To calculate the amount of sensible heat (H, in cal) involved in warming or cooling a substance through a specific temperature change, multiply the amount of matter (m, in grams) times the specific heat (c, in cal/g/C°) times the temperature change ($T_{higher} - T_{lower}$) in Celsius degrees.

$$H = m \times c \times (T_{higher} - T_{lower})$$

Based on the specific heat values shown in Table 1, how much heat is required to raise the temperature of three grams of each of the following substances from 15 °C to 25 °C? [Example: Determine (1) by multiplying 3 g of water × 1 cal/g/C° × (25 °C – 15°C)]

1. water: [(*3*)(*6*)(*18*)(*30*)] cal
2. sand: [(*3*)(*6*)(*18*)(*30*)] cal
3. alcohol: [(*3*)(*6*)(*18*)(*30*)] cal
4. marble: [(*3*)(*6*)(*18*)(*30*)] cal

Now consider equal amounts of heat added to equal masses in determining how much the temperature would be changed.

$$(T_{higher} - T_{lower}) = H/(m \times c)$$

Five cal of heat are added to 1 g of each of the following substances whose specific heats are given in Table 1. Each will be warmed by how many Celsius degrees? [Example: Item 5 - Determine temperature increase for 1 gram of sand by dividing 5 cal by 0.2 cal/g/C°.]

5. sand: [(*5*)(*10*)(*20*)(*25*)] C°
6. water: [(*5*)(*10*)(*20*)(*25*)] C°
7. ice   [(*5*)(*10*)(*20*)(*25*)] C°
8. water vapor [(*5*)(*10*)(*20*)(*25*)] C°

9. Five cal of heat are now removed from each of the substances described in Items 5 - 8 after they attained their final temperatures. All will end up with temperatures [(*lower than*)(*the same as*)(*higher than*)] their starting temperatures.

Latent Heat and Water

Latent heat is the amount of heat involved in changing the phase of a substance without an accompanying change in temperature. **Table 2** presents latent heat values (L) for water substance at one atmosphere of pressure and are expressed in calories per gram (cal/g). The amount of heat transferred during a phase change is: $H = m \times L$

10. Imagine a container filled with equal amounts of pure (fresh) water and crushed ice well mixed together. Its temperature will be 0 °C and will remain at that temperature until either complete freezing or melting occurs. According to Table 2, one gram of ice in the mixture will melt if [(*80*)(*540*)(*597*)(*677*)] cal of heat is added to the mixture.

11. If [(*80*)(*540*)(*597*)(*677*)] cal of heat is lost from the water/ice mix, one gram of water will freeze.

Table 2. Latent heat values

Fusion (*ice to liquid or liquid to ice*) at 0 °C — 80 cal/g

Sublimation (*ice to vapor*) or **Deposition** (*vapor to ice*) at 0 °C — 677 cal/g

Vaporization (*liquid to vapor*) or **Condensation** (*vapor to liquid*):

Temp. (°C)	0	10	20	30	40	50	100
Latent Heat (cal/g)	597	592	586	580	575	569	540

12. Table 2 shows that at 100 °C (boiling temperature of water at one atmosphere of pressure) the amount of heat absorbed when one gram of water vaporizes is [(*80*)(*540*)(*597*)(*677*)] cal. This amount is what is commonly reported as the value of the **latent heat of vaporization**.

13. However, liquid water can vaporize at any temperature. The energy involved in the phase change varies with temperature. Table 2 shows that as the temperature at which evaporation takes place increases, the amount of heat necessary to evaporate a gram of water [(*increases*)(*decreases*)(*remains the same*)].

14. Water vapor can condense at any temperature at which water exists as a liquid. The latent heat of condensation at any temperature is the same as the latent heat of vaporization at the same temperature. However, with condensation the latent heat is released to the environment rather than absorbed from the environment. According to Table 2, the value of the latent heats of vaporization and condensation at 10 °C is [(*80*)(*586*)(*592*)(*677*)] cal per gram.

15. Ice can vaporize without first melting (called sublimation) and vapor can deposit directly as ice (called deposition) at any temperature that ice exists. The table shows that for 1 gram of ice to directly change from solid to vapor at 0 °C requires [(*80*)(*540*)(*597*)(*677*)] cal.

16. Ice at 0 °C can also end up as vapor at 0 °C by first melting and then evaporating. One gram of ice at 0 °C requires the addition of 80 cal to melt. According to the table above, the resulting one gram of liquid water would require an additional [(*80*)(*540*)(*597*)(*677*)] cal to vaporize at 0 °C.

17. Whether ice sublimates at 0 °C or melts and then vaporizes at 0 °C, the total amount of energy absorbed in changing from solid to vapor is [(*80*)(*540*)(*597*)(*677*)] cal per gram.

Heat Energy Transfers

Imagine that on a sunny winter's morning a frozen water puddle initially at −5 °C warms to 0 °C, melts, the liquid warms to +10 °C, and disappears through evaporation. Determine the amount of heat required to change 10 grams of the ice initially at –5 °C to water vapor at +10 °C by completing **Table 3**. The first step in the process is shown as an example.

Complete the labeled blanks in Table 3 before answering Items 18, 19, and 20 below table.

Table 3. Heat energy absorbed as 10 g of –5 °C ice warms and changes phase to water vapor at 10°C

Step	Mass (g)	Heat Property? Specific Heat (cal/g/C°) or Latent Heat (cal/g)	Which? Sensible/ Latent Heat	Temperature Change (C°)	Heat Energy (cal)
Warm ice (–5 to 0 °C)	10	0.5 cal/g/C°	Sensible	5	25
Melt ice at 0 °C	10	80 cal/g	Latent Heat	none	*(18)* _____
Warm water (0 to +10 °C)	10	1 cal/g/C°	Sensible	10	*(19)* _____
Vaporize water at 10 °C	10	*(20)* _____ cal/g	Latent Heat	none	5920

18. [(***25***)(***800***)] cal   19. [(***10***)(***100***)] cal   20. [(***540***)(***592***)] cal/g

21. According to Table 3, the total amount of energy absorbed in the process of changing 10 g of ice at –5°C to vapor at 10 °C is [(***925***)(***5920***)(***6845***)] cal.

22. The processes described in the table involved heat transfer from the [(***environment to ice or water***)(*ice or water to the environment*)].

23. Which process required the addition of the greatest amount of heat? [(***warming ice***)(***melting ice***)(***warming water***)(***vaporizing water***)].

24. Based on the above completed table and the Law of Energy Conservation (Energy cannot be created or destroyed.), if 10 grams of water vapor at +10 °C condensed to water on the soil surface, cooled to 0 °C, froze, and then cooled to –5 °C, a total of [(***925***)(***5920***)(***6845***)] cal would be transferred from the water to the environment.

25. From our investigation of phase changes of water, it is evident that the global water cycle involves the flow of [(***energy***)(***mass***)(***both energy and mass***)] within Earth's climate system.

Summary: Water has a number of physical properties that cause it to play a central and dominant role in Earth's climate system. Of particular significance are the high specific heats and latent heats of water which were explored in this investigation. As we will see, these and other properties (including water vapor being the major "greenhouse gas") make water the core ingredient of climate, climate variability, and climate change.

Investigation 4B: WATER AND HEAT STORAGE AT THE EARTH'S SURFACE

Driving Question: *What role does water play in producing different seasonal temperature variations at places downwind of an ocean or large lake (maritime locations) versus places situated well inland (continental locations)?*

Educational Outcomes: To compare heat energy storage in water and land. To describe how energy from the Sun is absorbed and later released at land and water locations. To explain why ocean and large lakes impact weather and climate differently than land surfaces. To compare annual swings of surface air temperature at maritime and continental locations at the same latitude.

Objectives: Water's high specific heat compared to other Earth materials is one of the principal reasons why the ocean and large lakes impact weather and climate quite differently than do land surfaces.

Water has one of the highest specific heats (1 cal/g/C°) of all substances found in nature. The specific heats of solid Earth materials, such as limestone, marble and dry soil, are about one-fifth that of water. These differences in specific heats account for huge differences in the amounts of heat energy absorbed, stored, and released to the atmosphere by land and water which have important implications for weather and climate. Heat is the name given to energy transferred in response to a temperature difference within or between substances; heat is always transferred from where it is warmer to where it is colder.

After completing this investigation, you should be able to:

- Compare heat energy storage in water and soil.
- Compare seasonal changes in heat energy stored in water bodies versus land.
- Describe how climate is influenced by nearness to water bodies.

Heat Storage in Land and Water

The major purpose of this investigation is to determine the amounts of energy absorbed and later released at land and water locations at essentially the same latitude over the period of a year. During part of the year, lakes and the ocean are absorbers of huge quantities of energy delivered by the Sun. The rest of the year, these same lakes and ocean release the stored energy to the atmosphere above. Land surfaces absorb, store, and release energy, too, but in much different quantities.

Temperature measurements were obtained at two midlatitude locations at different depths in water (Lake Michigan) and in soil (St. Paul, MN). **Table 1** and **Table 2** include average annual maximum and minimum water and soil temperatures for different depth layers resulting from actual measurements made at the two locations, down to levels where temperatures remain steady throughout the year.

4B - 2

Table 1: Water
(Lake Michigan, Latitude 43°N)

Depth (cm)	Average Temperature (°C) Maximum (Summer)	Average Temperature (°C) Minimum (Winter)	Temperature Change (C°)	Volume (cm³)	Heat "Storage" (cal/cm³/C°)	Energy Stored (cal)
0 to 1000	24	4	*20*	1000	1	*20000*
1000 to 2000	19.5	4	1.	1000	1	5.
2000 to 3000	12.5	4	*8.5*	1000	1	*8500*
3000 to 4000	7.5	4	*3.5*	1000	1	*3500*
4000 to 6000	5.5	4	*1.5*	2000	1	*3000*
6000 to 11000	4.5	4	*0.5*	5000	1	*2500*
11000 to 15000	4	4	2.	4000	1	6.

"Total Energy Stored" in a one square-centimeter column of <u>water</u>: 7. _____ cal

Table 2: Land
(St. Paul, MN, Latitude 45°N)

Depth (cm)	Average Temperature (°C) Maximum (Summer)	Average Temperature (°C) Minimum (Winter)	Temperature Change (C°)	Volume (cm³)	"Heat Storage" (cal/cm³/C°)	Energy Stored (cal)
0 to 10	25.5	–6	*31.5*	10	0.5	*157.5*
10 to 100	23	–1.5	3.	90	0.5	8.
100 to 200	19.5	2	*17.5*	100	0.5	*875.0*
200 to 300	16	4.5	*11.5*	100	0.5	*575.0*
300 to 400	13.5	7	*6.5*	100	0.5	*325.0*
400 to 500	12.5	8.5	*4*	100	0.5	*200.0*
500 to 600	11	10	*1*	100	0.5	*50.0*
600 to 700	10.5	10	4.	100	0.5	9.
700 to 1300	9.5	9.5	*0*	600	0.5	10.

"Total Energy Stored" in a one square-centimeter column of <u>soil</u>: 11. _____ cal

Determine the temperature change between each layer's maximum and minimum temperatures over a year. Temperature Change calculations for most layers are already recorded as examples [*e.g. top layer - Water: 24 °C – 4 °C = 20 C° and Land: 25.5 °C – (-6 °C) = 31.5 C°*]. Complete the Temperature Change columns by calculating and recording missing values in the tables. Indicate your determinations below:

1. [(***0***)(***0.5***)(***15.5***)(***24.5***)]
2. [(***0***)(***0.5***)(***15.5***)(***24.5***)]
3. [(***0***)(***0.5***)(***15.5***)(***24.5***)]
4. [(***0***)(***0.5***)(***15.5***)(***24.5***)]

Imagine two vertical columns with cross-sectional areas of 1 cm², one of water and one of soil, at the above locations, and each extending downward to where temperatures remain steady throughout the year. With the information provided in the tables, it is possible to calculate the amount of heat energy stored and later released in individual layers of the two columns over the period of a year. Most of these calculations have already been done.

To calculate the amount of "Energy Stored"(**H**, in cal) involved in warming or cooling a given volume of a substance through a specific temperature change, multiply the volume of matter (**V**, in cm³) times the "heat storage" (**c′**, in cal/cm³/C°) times the temperature change ($T_{higher} - T_{lower}$) in Celsius degrees.

$$H = V \times c' \times (T_{higher} - T_{lower})$$

"Heat storage" describes heat necessary to change the temperature of a unit volume of a substance. [Note that this is different from specific heat because specific heat is based on temperature changes of a unit mass of a substance. As one gram of water occupies one cubic centimeter, it follows that one calorie of energy is equally required to change the temperature of one cubic centimeter of water one Celsius degree.]

In Table 1, the amount of heat stored in each layer of the water column was determined by multiplying the volume of water in the column layer times the **"heat storage"** value of 1 cal/cm³/C° times the temperature change. This was already done in the top layer having a volume of 1000 cm³ (1 cm x 1 cm x 1000 cm) that experienced a temperature change of 20 C°. Complete the Energy Stored column by calculating and recording missing values in the table. Indicate your determinations below:

5. [(***0***)(***25***)(***1102.5***)(***15500***)]
6. [(***0***)(***25***)(***1102.5***)(***15500***)]

7. Using the Table 1, Lake Michigan data, compute the **"Total Energy Stored"** in a vertical column of water with a cross-sectional area of one square centimeter extending from the surface of Lake Michigan to a depth of 15000 cm. It is found by adding together the energy stored in the individual layers of the column. Total Energy Stored is the amount of heat added to the water column as it warms from its average minimum temperatures in winter to its average maximum temperatures in summer. The **"Total Energy Stored"** in the Lake Michigan water column was [(***20000***)(***35000***)(***53000***)] cal.

In Table 2, the same procedure used in Table 1 is employed to calculate the Energy Stored in each layer of the soil column. It was assumed that 0.5 calorie of energy is required to change the temperature of one cubic centimeter of soil one Celsius degree. (This soil **"heat storage"** value is typical for a loam or silt loam soil with a water content of 25%.) Complete the Energy Stored columns by calculating and recording missing values in the tables. Indicate your determinations below:

8. [(*0*)(*25*)(*1102.5*)(*15500*)]
9. [(*0*)(*25*)(*1102.5*)(*15500*)]
10. [(*0*)(*25*)(*1102.5*)(*15500*)]

11. Using Table 2 data, compute the "**Total Energy Stored**" in a vertical column of soil with a cross-sectional area of one square centimeter at St. Paul extending from the land surface to a depth of 1300 cm. The "**Total Energy Stored**" in the soil column was [(*1103*)(*3310*) (*4387.5*)] cal.

It can be assumed that the energy entering and leaving the water and soil columns must pass through the top surfaces of the columns, their 1 cm^2 interfaces with the overlying atmosphere. Consequently, it is possible to contrast the heat storage capacities of land and water bodies to gain insight into their possible impacts on local and regional climate.

12. Comparing the amount of heat energy stored, the water column stored and later released [(*0.06*)(*16*)(*53*)] times as much heat energy as did the soil column.

13. Generalizing from the finding in Item 12, it appears that [(*soil*)(*water*)] has the capacity to store and later release much greater quantities of heat energy.

14. In Investigation 3A, incoming solar radiation data were presented that indicated in June and July, midlatitude locations receive on an average day approximately 500 cal of energy on every square centimeter (5 kWh/m^2) of horizontal surface. Assuming that Lake Michigan receives about the same amount of incident radiation (500 cal/cm^2) on an average summer's day, it appears that the Table 1 Lake Michigan water column absorbed an amount of heat energy equivalent to about [(*6.6*)(*16*)(*106*)] days of average summer incident solar radiation.

15. Assuming that St. Paul, MN received about the same amount of incident radiation (500 cal/cm^2) on an average summer's day, it appears that the Table 2 soil column absorbed an amount of heat energy equivalent to about [(*6.6*)(*16*)(*106*)] days of average summer incident solar radiation.

16. The stored heat is gradually released through Earth's surface to the atmosphere above during the colder portion of the year. Items 7 and 11 infer that this flow of energy would cause air temperatures above the [(*soil*)(*water*)] surface to be higher during this colder part of the year.

17. Prevailing winds carry the warmed air away from their heat sources. Consequently, locations downwind of [(*land*)(*water*)] will experience generally warmer winter climates.

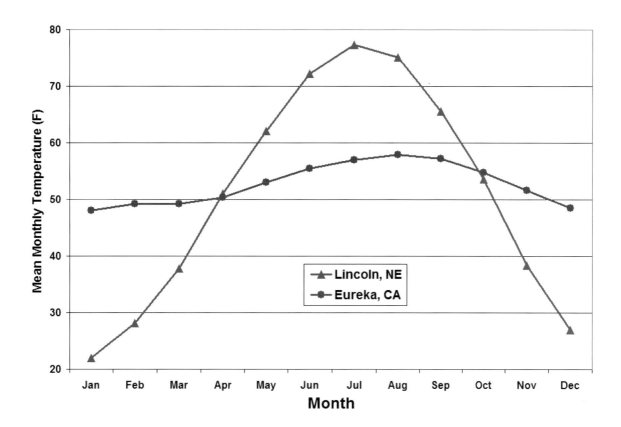

Figure 1.
Mean Monthly Temperature Curves for Lincoln, NE and Eureka, CA.

Examples of the annual swings of surface air temperature at maritime and continental locations are shown in **Figure 1**. The graph depicts mean (average) monthly temperature curves at two midlatitude locations, land-locked Lincoln, NE (Latitude 41° N) and coastal Eureka, CA (Latitude 41° N).

18. Compare the Lincoln and Eureka monthly mean temperature curves. The Eureka curve displays a [(*__greater__*)(*__lesser__*)] range (difference between highest and lowest) in monthly mean temperature than Lincoln.

19. Eureka's temperature peaks about one month [(*__earlier__*)(*__later__*)] than Lincoln. Eureka has an annual temperature curve typical of a maritime (downwind of an ocean or large lake) locality, whereas Lincoln exhibits an annual temperature curve that is characteristic of a continental (surrounded mostly by land) location.

20. This lowered warm season temperature and higher cold season temperature at coastal Eureka versus continental Lincoln [(*__is__*)(*__is not__*)] consistent with the heat storage behavior of water and land materials as shown by the data in Tables 1 and 2 above.

Summary: Water's high specific heat relative to other Earth materials is one of the principal reasons why the ocean and large lakes absorb huge quantities of energy delivered by the Sun compared to land at the same latitude. The differences in the amounts of heat energy absorbed, stored, and later released to the atmosphere by land and water have important implications for weather and climate. Air temperature is regulated to a considerable extent by the temperature of the surface which, in turn, is impacted by the amount of energy stored below the surface. Expansive land areas, with relatively little stored energy, produce **continental climates** with greater contrasts between average winter and summer temperatures. The **maritime climates** of places downwind of water bodies with greater amounts of stored energy experience much less contrast between average winter and summer temperatures.

Acknowledgment:

Soil temperature values were derived from data provided by the St. Paul Campus Climatological Observatory, Department of Soil, Water, & Climate, University of Minnesota, St. Paul, MN. Water temperature values were based on data provided by the Great Lakes Environmental Research Laboratory, National Oceanic and Atmospheric Administration, Ann Arbor, MI.

Investigation 5A: GLOBAL WATER CYCLE

Driving Question: *How does the global water cycle link the principal components of the Earth system?*

Educational Outcomes: To describe how unique properties of water place it in a major role establishing the boundary conditions and variability of climate. To explain how water cycles throughout the Earth system and impacts climate on all spatial and time scales, as well as how it transports heat globally as part of the never-ending drive towards a uniform distribution of energy.

Objectives: In the holistic Earth system perspective, Earth is composed of many interacting subsystems including the atmosphere, geosphere, hydrosphere, cryosphere, and biosphere. Earth's climate system encompasses aspects of these same components because their mutual interactions and responses to external influences determine climate on local, regional, and global scales.

After completing this investigation, you should be able to:

- Describe the components of the global water cycle within Earth's climate system.
- Explain ways in which the global water cycle links the various subsystems of Earth's climate system through flows of mass and energy.
- Explain the steady-state global water budget with more precipitation than evaporation occurring over land and more evaporation than precipitation taking place over ocean being balanced by the excess water on land dripping, seeping, and flowing back to sea.

The Global Water Cycle

In this investigation, some of the roles of water in shaping climate are explored, particularly as water participates in the complex redistribution of energy and mass by Earth's coupled atmosphere/ocean system.

Figure 1 is a depiction of the water cycle in terms of mass and mass flows between oceanic, terrestrial, atmospheric, and biospheric reservoirs. It provides a view of the water cycle already familiar to most people. What it does not display is the essential reason why water cycles through its reservoirs, namely, the accompanying energy flow. **The mass flow of water as portrayed in the figure is basically a response to the non-uniform distribution of energy in the Earth system.** Water's coexistence in three different phases, its high specific heat and latent heats, and the relative ease with which it changes phase within the temperature and pressure ranges on Earth, makes it the working fluid that absorbs, transports, and releases heat within the Earth system.

Water is the primary mover of energy from where there is relatively too much energy, to where there is too little energy. Winds transport water vapor to every location on Earth, including the highest mountain peaks. Changing back to liquid or solid within the

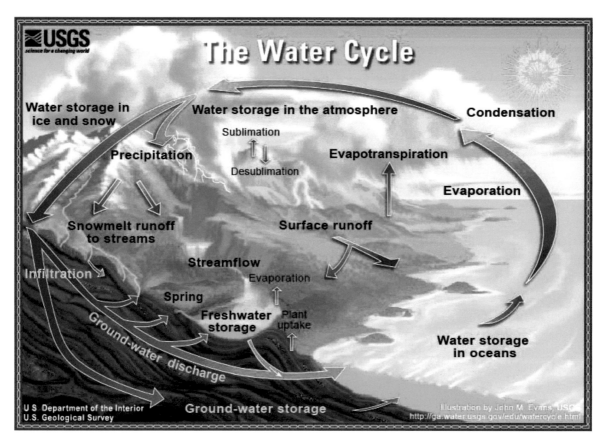

Figure 1.
The Water Cycle [USGS]

atmosphere, water begins its gravity-driven return trip to Earth's surface and eventually flows into the ocean. Ocean currents also transport energy as warm waters flow to higher latitudes and cold waters to lower latitudes. Development of a more complete and authentic depiction of the global water cycle, including both mass and energy flows, is an integral part of Earth's climate system.

Water flow in the atmosphere is the essential heat-driven *uphill* component of the water cycle that lifts water as vapor to great altitudes and transports it around the globe. The atmospheric water vapor flow embodied in the water cycle is invisible to our eyes because water vapor is transparent. However, special infrared sensors aboard weather satellites can detect the presence of water vapor (and clouds) in the atmosphere above altitudes of about 3000 m (10,000 ft).

You can view an animation of real Earth perspective full-disk GOES East water vapor imagery for the last day at: *http://www.ssec.wisc.edu/data/geo/index.php?satellite=east&channel=vis&coverage=fd&file=jpg&imgorani m=8&anim_method=jsani* (QR Code). At the SSEC Geostationary Satellite Images site, click on Imager Channel: Water Vapor 6.5 μm to view the most recent full-disk atmospheric water vapor.

QR Code

To examine an animated full global composite view covering the past week, go to: *http://www.ssec.wisc.edu/data/comp/wv/wvmoll.mpg*.

In the atmospheric water vapor imagery bright white patches represent the relatively cold tops of high clouds. Clouds occur where there are significant upward atmospheric motions. Medium gray regions depict mainly water vapor. These gray regions would probably appear clear on visible and ordinary infrared satellite images. Streaks and whirls of gray are common patterns in the water vapor images at the middle and higher latitudes. Dark areas are relatively dry portions of the middle atmosphere resulting from the sinking of low humidity air from higher altitudes.

1. Viewed in animation, the cloud patterns in the tropics and the curving swirls of water vapor in middle latitudes of the Northern and Southern Hemisphere portions of the images reveal that atmospheric motions **[(*do*)(*do not*)]** transport water vapor horizontally over great distances within the atmosphere.

2. Examine the general motion in the middle latitudes of the Northern and Southern Hemispheres. [To better analyze motions, consider using the control button to step forward through successive images.] In the middle latitudes, the dominant horizontal motion in both hemispheres is from **[(*higher to lower latitude*)(*east to west*) (*west to east*)]**.

3. Examine the motion seen in the tropical latitudes. The numerous bright white cloud patches at these latitudes signify upward convection currents and thunderstorms. These tropical clouds embedded in air with relatively high water vapor concentrations generally mark the ***Intertropical Convergence Zone (ITCZ)***. Here, intense solar energy arriving at Earth's surface heats near-surface air, promoting evaporation (and transpiration, especially in tropical rainforests) and upward vertical motions. The dominant horizontal motion in the ITCZ is from **[(*east to west*)(*west to east*)]**.

Figure 2 displays two composite images acquired one day apart. The upper image is from 0000 UTC 27 DEC 2011 (7:00 pm EST on the 26th) and the lower image is 24 hours later, 0000 UTC 28 DEC 2011. During this time period a winter storm system delivering considerable precipitation traveled up the U.S. East Coast packing strong winds that downed trees and power lines leading to delays in post-holiday travel.

4. Note on the upper image the broad band of light gray extending from the tropical Pacific Ocean across central Mexico to the Southeastern states. The brightest area in that band centered northwest of Florida corresponds to the major winter storm that brought high winds and heavy rains across a broad area of the country while traveling up the East Coast. Such midlatitude swirls are storm systems that transport warm humid air poleward to be replaced by colder drier polar air moving equatorward. Such storm motions transfer **[(*water mass*)(*heat energy*)(*both*)]** within the Earth system.

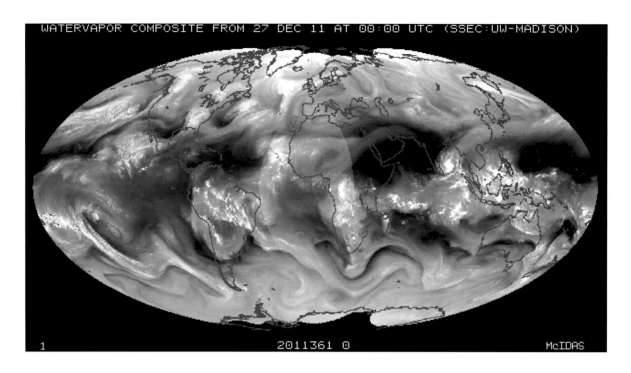

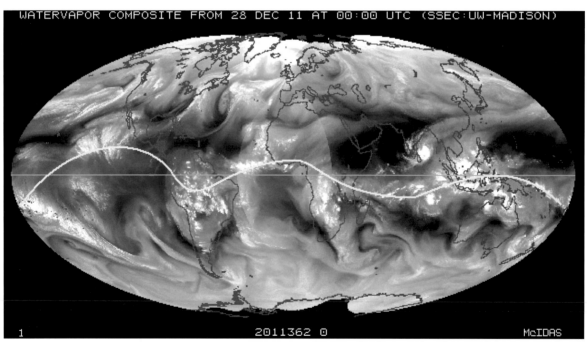

Figure 2.
Composite satellite images displaying the distribution of water vapor in the middle levels of the atmosphere at 0000 UTC 27 December 2011 (upper view) and 0000 UTC 28 December 2011 (lower view). The lower view has lines signifying the positions of the equator (green) and average December position of the Intertropical Convergence Zone (ITCZ) in yellow. [SSEC-UW]

5. Based on what you learned about midlatitude atmospheric motions (generally west to east) from the animated images, it appears that the plume of water vapor fueling the winter storm system described above originated primarily (upper image) from the

[(*Pacific Ocean*)(***Atlantic Ocean***)]. It is evident from a global water cycle perspective, that water evaporated from ocean surfaces was incorporated in the overlying air and transported to areas above land surfaces where significant quantities then precipitated as rain, freezing rain (icing), and snow.

6. Now examine the lower image in Figure 2 showing atmospheric water vapor distribution one day after the upper image. Focus in on the U.S. East Coast area. It is evident from the two images that the winter storm had moved generally towards the [(***east***)(*west*)]. This movement, combined with north or south components, is commonly characteristic of midlatitude storm systems. The patterns and frequency of such storm systems have major impacts on local and regional climate at places in their paths.

7. A straight green line has been drawn across the midsection of the lower image in Figure 2 to represent the equator. The yellow curve approximates the center of the average December position of the Intertropical Convergence Zone (ITCZ). The ITCZ is a relatively narrow, irregular band of convective clouds and thunderstorms roughly parallel to the equator in which huge quantities of water vapor enter the atmosphere from underlying warm surface ocean waters and tropical rainforests (e.g., Amazon Basin). Note the general agreement between the brightest clouds surrounded by light gray shadings and the average December ITCZ position. They show that at this time of the year the ITCZ is largely [(*north*)(***south***)] of the equator. Six months later, the ITCZ is found almost entirely on the opposite side of the equator.

8. In Figure 2, two darker irregular bands surrounding Earth and located north and south of the equator indicate the presence of [(***dry***)(*humid*)] air. These are broad regions of relatively high surface air pressure and associated sinking air which produce relatively persistent clear skies and fair-weather conditions, which in turn have major impacts on local and regional climates.

9. The flow of atmospheric water vapor is a major energy transport mechanism in Earth's climate system. The energy flow starts with the evaporation of water from ocean and land surfaces or transpiration by vegetation transferring energy
[(*from air to water and land surfaces or vegetation*)
(***from water and land surfaces or vegetation to air***)].

10. Atmospheric water vapor is often carried great distances before it condenses to liquid or solid cloud particles, thereby transporting huge quantities of [(*sensible*)(***latent***)] heat energy from one place to another.

11. Assume that much of the heat energy added to the environment by the late December 2011 Southeastern U.S. winter storm depicted in Figure 2 was carried in the water vapor plume shown flowing generally northeastward into the storm. The heat energy carried by the plume in the lower image was now originating most directly from the [(***Pacific Ocean***)(*Atlantic Ocean*)]. In combination with Item 5, it is evident that the global water cycle embodies both mass and energy flow.

The global water cycle's transport of mass (and, indirectly, energy) can be traced via NASA's Tropical Rainfall Measuring Mission's (TRMM) rainfall animations. TRMM provides estimates of rainfall between about 40 degrees N and 40 degrees S using sensors on a satellite.

Go to: *http://trmm.gsfc.nasa.gov/publications_dir/global.html* and click on A WEEK of Rainfall "Medium Quicktime" or "Medium Mpeg" to observe rainfall patterns for the previous seven days.

QR Code

12. The TRMM Rainfall patterns during the past week, when generalized, show evidence of [(***passage of storm systems from ocean to land and from land to ocean***) (***general eastward motion of midlatitude storms***) (***general westward motion of tropical storms***)(***all of these***)].

At the same NASA TRMM website, click on A WEEK of Rainfall ACCUMULATION "Medium Quicktime" or "Medium Mpeg" to observe accumulated rainfall patterns for the previous seven days.

13. Broad areas in the TRMM global accumulated rainfall pattern for the past week that exhibited evidence of little or no rainfall generally [(***show no relationship to***) (***coincide with***)] the dark areas in the animated global views of atmospheric water vapor you viewed at the beginning of this investigation. As you recall, the dark areas denoted low water vapor content.

Summary: Water cycles throughout the Earth system. Water vapor flow in the atmosphere is the essential heat-driven ***uphill*** component of the water cycle that lifts water as vapor to great altitudes and transports it around the globe. The atmospheric energy flow embodied in the water cycle is largely invisible to our eyes but special infrared sensors aboard weather satellites enable us to monitor it. The atmosphere delivers water vapor globally and impacts climate on all spatial and time scales, acting as a major agent transporting heat in a never-ending drive towards a more uniform distribution of energy.

Investigation 5B: WATER VAPOR FLUX AND TOPOGRAPHICAL RELIEF

Driving Question: *Do precipitation amounts on windward and leeward sides of mountains differ?*

Educational Outcomes: To investigate the role of elevation as a boundary condition of climate. To examine the influence of topographical relief on precipitation. To explain the differences in precipitation patterns on the windward and leeward slopes of mountain ranges.

Objectives: The flow of air in the atmosphere forced vertically by changes in the elevation of Earth's surface plays major roles in determining climates, particularly in regions exhibiting significant topographical relief. After completing this investigation, you should be able to:

- Compare precipitation amounts at locations on the windward and leeward sides of a mountain range.
- Explain how and why precipitation totals vary on the windward and leeward slopes of a mountain range.
- Describe the implications of topographically-induced variations in precipitation for activities that depend on the fresh water supply.

Orographic Precipitation

Vertical motions in the atmosphere forced by changes in the elevation of Earth's surface play major roles in determining climates, particularly in regions exhibiting significant topographical relief. *Orographic* lifting is an important boundary-condition mechanism that can bring air to saturation and subsequently cause cloud formation and precipitation. Air flowing up the windward slopes of a mountain range is subjected to decreasing air pressure, which brings on expansion and cooling. With sufficient ascent and cooling, the air becomes saturated, clouds form, and precipitation usually develops. Air flowing down the leeward slopes of a mountain range is compressed by the increased air pressure and warms. With warming, existing clouds vaporize in the descending air. Once the clouds have vaporized, the relative humidity decreases as the sinking air continues to warm by compression.

Where large-scale winds blow persistently toward a prominent mountain range, the climate of the windward slope can be much wetter than the climate of the leeward slopes. And, in fact, the region of reduced cloudiness and precipitation downwind may extend great distances from the mountain range. In this so-called *rain shadow*, meager precipitation coupled with frequent sunny skies that promote evaporation means agricultural activity is likely to be limited without irrigation.

1. The topographic influence on precipitation is especially evident across the U.S. Pacific Northwest where the large-scale atmospheric circulation across the region is fairly consistent. For a view of the recent flow of water vapor over that region, go to:

http://www.wrh.noaa.gov/satellite/. Scroll down to the table and its "Water Vapor" section. In its "Western US" coverage, click on the 16-km resolution "Animation." The general flow of water vapor at the U.S. Pacific Northwest latitudes is towards the [(*west*)(*east*)].

Prevailing winds deliver relatively humid air masses from their Pacific Ocean source regions. North-south oriented mountain ranges provide barriers to the humid flow, forcing it upward or around the topographical obstacles. In winter the region's weather is often dominated by a procession of storm systems from over the Pacific Ocean. In summer, storm systems are less frequent and accompanied by less precipitation.

Figure 1 below is a colorized relief map of Washington State and adjacent areas with near sea-level elevations of the land in green. North is to the top of the map. The Cascade Mountains trend north-south through the center of Washington with the Olympic Mountains in the west and the Coast Range to the southwest into Oregon.

Figure 1.
Relief map of Washington State. [Modified from map by Ray Sterner, Johns Hopkins University Applied Physics Laboratory]

Numbers plotted on the map identify locations listed in **Table 1** below. Table 1 gives the elevation (in feet above mean sea level), and mean seasonal and annual precipitation (in inches of rain plus liquid equivalence of snow) for eight stations. The seasons are defined by meteorological convention: winter (December - February), spring (March - May), summer (June - August), and autumn (September - November). Examine the climatic information in the Table.

Table 1. Mean seasonal and annual precipitation at selected localities in Washington State and at Astoria, OR, and Lewiston, ID.

Location	Elevation (feet)	Winter (inches)	Spring (inches)	Summer (inches)	Autumn (inches)	Annual (inches)
1. Astoria, OR	6	28.4	14.9	5.0	19.3	67.4
2. Centralia	183	19.0	9.7	3.6	12.7	45.1
3. Longmire	2759	28.3	17.4	7.3	23.9	77.2
4. Mt. Rainier	5426	45.7	23.6	8.6	31.7	109.6
5. Yakima	1063	3.4	1.7	1.2	1.9	8.2
6. Richland	370	2.7	1.7	1.0	1.8	7.2
7. Walla Walla	948	5.5	4.2	1.9	4.4	16.0
8. Lewiston, ID	1433	3.3	3.5	2.7	3.1	12.6

2. At virtually all stations, the [(*winter*)(*summer*)] season receives the greater amount of precipitation on average.

3. As mentioned earlier in this investigation, greater precipitation at that season of the year is the consequence of [(*more*)(*less*)] frequent storms tracking inland from the Pacific Ocean.

Average annual, winter, and summer precipitation amounts, along with elevations, for each of the stations in Table 1 are plotted in **Figure 2** below. The relative locations of the stations in the general direction of west at the left and east to the right are given along the base. Elevations of the stations, in feet above mean sea level, are depicted with the vertical altitude scale shown to the lower right. Precipitation amounts for each station are plotted in the upper portion of the figure (vertical scale of precipitation in inches shown at the left). The magenta solid line connects triangles showing the annual mean precipitation by station, the aqua short-dash line connects diamonds indicating winter mean precipitation, and the red long-dash line connects squares signifying summer mean precipitation.

4. Traveling eastward from Astoria (1) on the Pacific coast to Centralia (2), average annual, winter, and summer precipitation [(*increases*)(*decreases*)] with little change in elevation but with increasing distance from the Pacific Ocean source region of moisture.

5. From Centralia eastward to Mt. Rainier (4, at the Paradise Ranger Station) the elevation of the Earth's surface rises about [(*52*)(*520*)(*5200*)] feet.

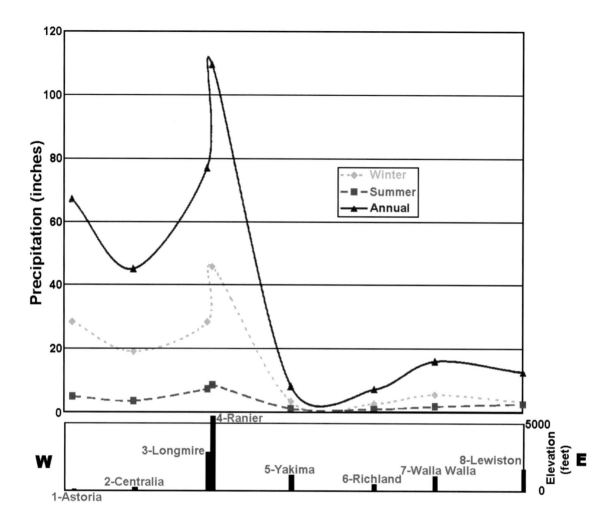

Figure 2.
Average annual, winter, and summer precipitation amounts, along with station elevations from west (W) to east (E) of locations shown on Figure 1 relief map.

6. From Centralia to the Mt. Rainier Ranger Station, average annual precipitation [(*increases*)(*decreases*)] with increasing elevation.

7. This is occurring on the mountain range's [(*windward*)(*leeward*)] slope.

8. From the two seasonal curves shown, greater precipitation occurs during [(*summer*)(*winter*)].

9. From the Mt. Rainier Paradise Ranger Station eastward to Yakima (5), the elevation drops about 4400 feet and the average annual precipitation [(*declines*)(*increases*)] dramatically.

10. On an average annual basis, Yakima receives about [(*7.5*)(*75*)(*750*)] percent of the precipitation at the Mt. Rainier Paradise Ranger Station. Yakima is located on the leeward side of Washington State's Cascade Mountain range, of which Mt. Rainier is the peak.

11. Compare the two segments of the average annual precipitation curve west and east of Rainier's location. The segment [(*west*)(*east*)] of Mt. Rainier indicates noticeably less precipitation as compared to the other segment as is characteristic of rain shadows caused by topographical barriers.

12. Further comparison of the two average annual precipitation curve segments indicates this rain shadow extends to about [(*Richland*)(*Walla Walla*)(*Lewiston and beyond*)].

13. Based on the two average annual precipitation curve segments and knowing clouds are required for precipitation, residents to the [(*west*)(*east*)] of Mt. Rainier are likely to experience the greater number of cloudy days through the course of a year.

14. The summer and winter precipitation curves indicate these same residents are more likely to have cloudy skies in [(*summer*)(*winter*)].

15. The indigenous vegetation surrounding Yakima and Richland (6) is likely to be [(*rainforest*)(*desert*)].

16. On the basis of average annual precipitation, irrigation will probably be most needed for agriculture at [(*Astoria*)(*Longmire*)(*Richland*)].

17. Average air temperature usually declines with increasing elevation. In traveling from Centralia eastward to Rainier, you would expect [(*falling*)(*rising*)] average temperatures.

18. Another reason for greater average precipitation at higher elevations (besides orographic lifting) is the fact that highlands are closer to the base of precipitation-producing clouds than lowlands. In general, the closer the ground is to cloud base, the [(*less*)(*more*)] the amount of precipitation that can vaporize in the unsaturated air beneath the clouds to reduce actual precipitation amounts.

Summary: Vertical motions in the atmosphere forced by topographical relief can be a major boundary condition in determining climates on local and regional scales. Orographic lifting can bring air to saturation, the formation of clouds and possibly precipitation, resulting in moist climates. Continued flow of the same air descending on leeward sides of mountains can produce relatively cloud-free and low humidity conditions with little precipitation. The resulting "rain shadows" can extend far downwind where semi-arid conditions prevail and limit agricultural productivity unless augmented by irrigation.

GLOBAL ATMOSPHERIC CIRCULATION

Driving Question: *What are the principal features and causes of Earth's planetary-scale atmospheric circulation?*

Educational Outcomes: To describe the planetary-scale circulation that arises from the combined effects of Earth's directional receipt of solar energy and Earth's rotation. To explain the Coriolis Effect that arises when air (and other objects) moves over a rotating Earth.

Objectives: After completing this investigation, you should be able to:

- Describe the time-averaged global circulation of Earth's atmosphere.
- Explain the Coriolis Effect, the deflection of moving air (as well as water and other objects) traveling horizontally in the Earth system.
- Explain in general terms how the planetary atmospheric circulation arises from the combined effects of the directional receipt of solar radiation and Earth's rotation.

General Circulation of Earth's Atmosphere

In its simplest terms, the general circulation of Earth's atmosphere arises from the combined effects of (a) the directional receipt of solar energy from space, (b) Earth's rotation, and (c) character of Earth's surface (e.g., topographical relief, land/water surfaces) as the Earth system works to maintain planetary energy equilibrium with surrounding space.

The solar energy that fuels Earth's climate system is characterized by both magnitude and direction. Although the Sun appears as a tiny object in our daytime sky, its energy arrives in essentially parallel rays perpetually striking half of the rotating and revolving spherical Earth. In **Figure 1**, the solid curve denotes the long-term average rates of the absorbed solar radiation received on Earth by latitude. Simultaneously, Earth is a continuous radiator of infrared radiation to space from everywhere on the planet. The dashed curve describes the average rates of infrared radiation emitted to space by latitude. The figure is scaled so the total areas under each of the curves are essentially equal and are directly proportional to the total amounts of energy received or emitted respectively. This is because over time Planet Earth must be in radiative equilibrium with space, losing as much energy to space as it receives from space.

As seen in Figure 1, Earth's lower latitudes receive greater radiational heating from the Sun than they lose to space by emitted infrared radiation (surplus area) over the period of a year. At higher latitudes, more energy is lost to space as infrared radiation than absorbed from incoming solar radiation (deficit area). However, the lower latitudes are not continually warming year after year, nor are the higher latitudes becoming progressively cooler because of these imbalances in radiational heating and cooling. That is, the meridional (equator-to-pole) temperature gradients tend to stabilize in the Northern and Southern Hemisphere

Figure 1.
Variation by latitude of rates of absorbed solar radiation and outgoing infrared radiation derived from satellite sensor measurements. [From NOAA/NESDIS]

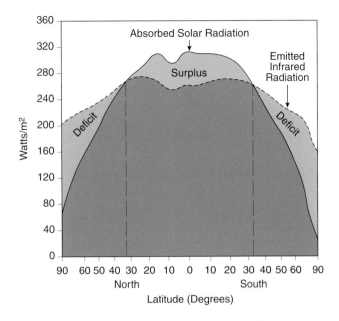

because heat energy is transported poleward from low latitudes (surplus areas) to higher latitudes (deficit areas).

Global energy balance requires that the surplus of incoming energy in the lower latitudes be transferred to the higher latitudes where losses of infrared radiation to space dominate. The fluid atmosphere and ocean provide the mechanisms that transport heat energy poleward.

1. According to Figure 1, when averaged over a year, the rate of absorbed incoming solar radiation equals the rate of outgoing infrared radiation near [(*16°*)(*23°*)(*32°*)(*46°*)] N and S latitudes.

2. Heat energy in the energy "surplus" latitudes must flow to the energy "deficit" latitudes in order to stabilize the equator-to-pole temperature gradients. The latitudes marking the boundaries between these "surplus" and "deficit" regions must be where the greatest rates of poleward heat energy transport take place. Figure 1 shows the latitude(s) where poleward heat energy transport is greatest is (are) near [(*0°*)(*23° N/23° S*)(*32° N/32° S*)(*46° N/46° S*)] when averaged over a year.

Figure 2 is presented to explore the roles played by receipt of solar energy and Earth's rotation in determining the global atmospheric circulation.

Figure 2(a) represents a model of an imaginary planet about the size of Earth that has the same side continuously facing the Sun. The planet has a smooth solid surface and an atmosphere. The straight (orange colored) arrows represent essentially parallel rays of sunlight. The amount of solar radiation received is greatest where the Sun's rays are perpendicular to the planet's surface (the place on the planet where the sun would be directly overhead, called the sub-solar point) and decreases to zero intensity where the Sun's rays graze (are tangent to) the planet's surface. Half of the celestial object is in perpetual darkness.

a. No Rotation

Figure 2.
Models of planets with solid smooth surfaces and an atmosphere, having (a) no rotation, (b) relatively slow rotation, and (c) fast rotation. The axis of rotation in (b) and (c) is oriented perpendicular to Sun's rays. [Adapted from F.W. Taylor, *Elementary Climate Physics*, Oxford University Press, 2005]

b. Slow Rotation

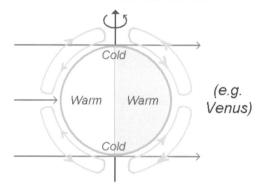

c. Fast Rotation

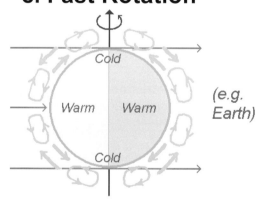

3. Atmospheric motion on this Figure 2(a) non-rotating planet is depicted in cross-section by the blue lines with arrowheads indicating direction. The most intense heating of the atmosphere takes place on the planet's surface at the sub-solar point, where the air expands and becomes less dense. At that location, the warmed air [(*sinks*)(*rises*)]. At the anti-solar point (on the opposite side of the planet from the sub-solar point), the temperature is lowest, the air is most dense, and the air sinks.

The atmosphere responds to the differential heating due to the directional receipt of solar energy. Resulting differences in air density produce forces that put the atmosphere into motion. A giant convection cell develops to transport cold air from the anti-solar point along the surface to the sunlit side where it is heated by incoming solar radiation. It rises in the region of the sub-solar point, travels at higher altitudes to the night side of the planet, and sinks in the region of the anti-solar point. Imagine what is depicted in Figure 2(a) in three dimensions. The coldest surface air on the planet at the place opposite the Sun's position spreads over the surface and advances to the sunlit side. There it is heated by the Sun's rays as it progresses to the sub-solar point where it is warmest. There it mushrooms upward and spreads outward at higher altitudes as it begins its trek to the dark side of the planet. On the dark side, it cools as it converges toward the anti-solar point. There, it sinks and begins its surface flow back to the sunlit side as it completes its planetary-scale circulation pattern.

4. A figure or object is symmetrical if an axis or imaginary line exists about which an object could be rotated and still resemble its original shape. In the three-dimensional view of Figure 2(a), the convection circulation pattern shown would be seen as symmetrical about a straight line [(*perpendicular to the Sun's rays*) (*drawn from the center of the Sun through the sub-solar and anti-solar points*)].

5. **Figure 2(b)** introduces planetary rotation around an axis oriented perpendicular to the Sun's rays. Actual observations of planetary atmospheres have shown that even with slow rotation, as depicted by Figure 2(b), the global atmospheric circulation will be symmetrical to [(*a line drawn from the center of the Sun through the sub-solar and anti-solar points*)(*the axis of rotation*)]. The Planet Venus, which rotates only once in 243 Earth days, spins rapidly enough to distribute heat adequately throughout its equatorial region to cause the rotation poles to become the coldest places on the planet.

6. The relatively slow rotation of the planet is adequate to cause the axis of symmetry for its atmosphere's general circulation to shift from its Figure 2(a) position until it coincides with the axis of rotation. With the coldest air at the poles and the warmest air along the equator, surface winds flow from the [(*poles to the equator*)(*equator to the poles*)]. In a three-dimensional view with North Pole at the top, the convection circulation pattern shown in Figure 2(b) would be seen as two single Hadley cells; one in the Northern Hemisphere and one in the Southern Hemisphere.

Figure 2(c) presents a planet (think Earth) with relatively fast rotation (e.g., one rotation in one Earth day). Due to the faster rotation, the hemisphere-scale Hadley cells described in Figure 2(b) are no longer stable. With the exception of air in motion at the equator, air moving freely across the surface of the rotating planet will exhibit motion along a curved path. This change in direction is observed because the motion is being measured relative to a rotating frame of reference (i.e., the planet's surface). This departure from straight-line motion is referred to as the **Coriolis Effect**.

7. Caused by planetary rotation, the Coriolis Effect is absent at the equator but increases as latitude increases, reaching a maximum at the poles. Its direction of action, to the right or left, is determined by the sense of planetary rotation. In Figure 2(c), imagine

yourself in space looking down on the planet's north pole. The planet would be seen rotating counter-clockwise, as shown by the curved black arrow. Now imagine yourself looking down on the planet's south pole. From above the south pole, the planet would be seen to be rotating [(*counter-clockwise*)(*clockwise*)]. Applied to Earth, these opposite senses of rotation cause air moving horizontally to turn towards the right in the Northern Hemisphere and toward the left in the Southern Hemisphere.

8. Figure 2(c) shows that the intense heating in the tropics initiates rising air currents as expected in a Hadley cell circulation. At upper levels the air moves to higher latitudes while being deflected to the right (Northern Hemisphere) or left (Southern Hemisphere). By about 30 degrees N or S, the upper-level air would be flowing either east (Northern Hemisphere) or west (Southern Hemisphere) and essentially "piling up." This added air raises surface air pressure, which pushes air near the surface towards both lower and higher latitudes. The flows toward the equator produce the **Trade Winds** and those towards higher latitudes contribute to the **Prevailing Westerlies**. In the Northern Hemisphere, the south-moving surface air is deflected to the right by Earth's rotation. Wind direction is named according to the direction from which the wind blows so that these Northern Hemisphere flows are called the [(*Northeast Trades*)(*Southeast Trades*)].

9. On the fast rotating planet (again, think Earth), the Hadley cells are confined to the lower latitudes. They are bounded at about 30 degrees N and S by generally [(*cloudy*)(*clear-sky*)] semi-permanent high pressure areas characterized by large-scale sinking air that is part of the Hadley-cell circulation.

10. Averaged over time, atmospheric circulation in the middle latitudes reveals little, if any, detectable Hadley cellular structure. Instead, it is somewhat chaotic with waves, eddies (irregular whirls in thermal and mechanically induced turbulent flow), and storm systems (e.g., extratropical cyclones). Middle latitudes include the region where, according to Figure 1, the [(*minimum*)(*maximum*)] rate of horizontal (poleward) heat energy transport takes place. Scientists have found that the combination of cellular and turbulent eddy energy transfer observed in the middle latitudes moves energy at the fastest possible rate.

11. At the highest latitudes a **Polar cell**, approximately centered on the pole, is thermally maintained. Air masses migrating from lower latitudes are sufficiently warm and moist to undergo convection at about 60 degrees N and S, where rising air flows poleward, cools, and descends. The cold, dense air produces high surface air pressure. This air pushes towards lower latitudes. In the Arctic, the Coriolis Effect deflects the southerly flow to the right. In the Antarctic, the Coriolis Effect deflects the northerly flow to the left. Whether in the Northern Hemisphere or the Southern Hemisphere, the surface circulation of the polar cells becomes the [(*Polar Westerlies*)(*Polar Easterlies*)].

Circulation of Earth's Real Atmosphere

Striking evidence of the general circulation of the atmosphere can be seen via animated satellite views of Earth. Go to the following animation of full-disk views of Earth as seen from the European METEOSAT geostationary weather satellite located above the Gulf of Guinea: *http://www.gerhards.net/astro/wolken_200705_en.html*.

12. Scroll down to the global view and click on the play button to the lower left of the image to start the animation. The view is centered on Earth's surface directly under the satellite at the intersection of the equator (0° Latitude) and the Prime Meridian (0°Longitude). This animation shows Earth's cloud circulation during a full month (May 2007) as seen in EUMTSAT infrared imagery. The bright, blotchy clouds in the equatorial region can be seen flowing generally from [(*west to east*)(*east to west*)]. (Clouds move with the horizontal wind and so their motions are an indicator of wind direction.)

13. This band of bright clouds in the equatorial region is an indication of [(*sinking*)(*rising*)] air. The band generally marks the locations where surface winds in the Northern and Southern Hemispheres blowing towards the equator meet. Called the ITCZ (Intertropical Convergence Zone), the band of cloudiness is characterized by the bright clouds indicating intense thunderstorm activity and by high rainfall rates. It roughly parallels the equator around the globe.

14. The animation shows that to the north and south of the ITCZ are broad expanses of relatively clear skies (e.g., Sahara Desert of northern Africa and the Kalahari Desert of southern Africa). Similar extensive clear sky areas are found around the globe at about the same northern and southern latitudes. The general absence of clouds implies that at these latitudes, the atmosphere exhibits persistent [(*sinking*)(*rising*)] air motion.

15. Note the wisps of less bright clouds that are frequently seen spewing towards higher latitudes from the ITCZ cloud complexes. They are high-altitude clouds moving towards the [(*northwest in the Northern Hemisphere and towards the southwest in the Southern Hemisphere*)(*northeast in the Northern Hemisphere and towards the southeast in the Southern Hemisphere*)]. Their motions are consistent with global circulation models that depict low-latitude circulations resembling convection currents between the ITCZ and about 30 degrees N and S. These are the **Hadley cells**. The surface components of these cells are the Northeast Trade winds in the Northern Hemisphere and the Southeast Trades in the Southern Hemisphere.

16. Further examination of the animation shows that in the middle latitudes poleward of the relatively clear areas in both the Northern and Southern Hemispheres are swirls of clouds marching generally from [(*west to east*)(*east to west*)]. These cloud motions also show a tendency for movement towards higher latitudes.

The satellite images we have been examining in this animation show only about 42% of Earth's surface due to the satellite being positioned in space at an altitude of about 36,000 km

(22,300 mi) above Earth's surface. Geostationary satellites are not positioned far enough from Earth to provide views of Earth's surface and atmosphere at the highest latitudes. Consequently, the animation being examined does not show atmospheric circulations in polar regions. However, it does provide considerable visual evidence of the general circulation of Earth's atmosphere.

[Similar recent full-disc animations of geostationary satellite views from around the world ending with the latest imagery available can be found via: *http://www.ssec.wisc.edu/data/geo/*. At this website, at the left, click on "8 Image Animation" button under "Latest Image/Animation", and under "Geographic Coverage", click on "Full Disk" button.]

Global Views of Earth's General Circulation: To view the latest ten-day global montage movie which includes cloud-top satellite infrared imagery, go to either:
  (AniS Java format) - *http://www.ssec.wisc.edu/data/comp/cmoll/cmoll.html*
  (MPEG format) - *http://www.ssec.wisc.edu/data/comp/cmoll/cmoll.mp4*

Updated every six hours, the global montage contains cloud images that are a combination of GMS, GOES-8, and Meteosat imagery. Sea surface and synoptic observation temperatures over land are also presented in the montage. The AniS Java format enables you to set the animation speed bar above the image. In the MPEG format version, you can control animation speed by repeated clicking on the forward button at the lower center of the image.

17. During the period of the global montage animation you are viewing, clouds in Earth's tropical region show a worldwide general motion that trends towards the [(*east*)(*west*)].

18. The swirls of clouds at the middle and higher latitudes show a worldwide general motion that trends towards the [(*east*)(*west*)].

Note the daily changes in temperatures over land surfaces. The temperatures are coded as shown in the "SYNOPTIC OBS" portion of the color bar under the montage. The time is reported in the bottom line of data in Coordinated Universal Time (UTC).

19. Observe daily temperature changes at a particular location by focusing on Ghana in western Africa (located close to the equator and on the Prime Meridian). There, at 0° Longitude, the local time for 00 UTC is midnight, 06 UTC is 6 am (about sunrise), 12 UTC is 12 noon, and 18 UTC is 6 pm (about sunset). Watching several daily cycles, the minimal daily temperatures seen on the sequence of images is at [(*midnight*)(*6 am*)(*noon*)(*6 pm*)] local time.

20. Now, over several daily cycles, note the longitudinal change in the temperatures of land surfaces as a day progresses. The locations of highest land-surface temperatures (the regions of deepest red color) migrate from [(*east to west*)(*west to east*)].

Summary: The planetary-scale circulation of Earth arises from the combined effects of Earth's directional receipt of solar energy and Earth's rotation. The daily temperature cycles observed in Items 19 and 20 result from the combined effects of the receipt of solar energy and Earth's rotation. Being close to the Equator, Ghana experiences nearly equal periods of daylight and darkness that produce a fairly uniform daily temperature range throughout the year. The migration of daily high temperatures observed in Item 20 results from the eastward rotation of Earth. These are aspects of Earth's directional receipt of solar energy and Earth's rotation that are the primary boundary conditions that determine Earth's climate system.

Investigation 6B:

GLOBAL ATMOSPHERIC CIRCULATION – ROSSBY WAVES

Driving Question: *What are the upper-air atmospheric long waves, called Rossby waves, that exist in the planetary westerlies and how are they related to weather and climate?*

Educational Outcomes: To describe the upper-air long-wave circulation in the middle latitudes, called Rossby waves, which arises as part of the global atmospheric circulation. To explain its relationship to surface weather features and the impact of their occurrence on local and regional climate.

Objectives: After completing this investigation, you should be able to:

- Describe the planetary-scale upper-air flow pattern called Rossby waves.
- Demonstrate how Rossby waves are detected by interpreting radiosonde-based data displayed on an upper-air constant pressure map.
- Explain the general relationships between Rossby wave ridges and troughs, with surface weather and climate.

Upper-air Global-scale Atmospheric Circulation

The middle latitude westerlies of the global atmospheric circulation merit special attention as those in the Northern Hemisphere govern the weather and climate over much of North America. Surface and near-surface atmospheric motions in the prevailing westerlies are dominated by gigantic eddies and whirls (roughly circular movements as seen from above), within the larger main horizontal air currents. These are detected on Northern Hemisphere surface weather maps as clockwise circulations within areas of fair-weather high pressure systems (Highs or anticyclones) and counterclockwise motions in areas of storm-prone low pressure systems (Lows or cyclones). (The circulations in these Highs and Lows are reversed in the Southern Hemisphere.) In the middle and upper reaches of the troposphere, these circulations give way to a meandering long-wave flow stretching around the planet. This series of upper-air, planetary-scale undulating waves is called a **Rossby wave** pattern. Together, the near-surface weather systems and upper-air flows marked by Rossby waves provide transport mechanisms that move heat energy poleward.

Rossby Waves: The emergence of cold air masses from the high-latitude polar cells is instrumental in generating the Rossby waves in the middle and higher latitudes. **Figure 1** is the Northern Hemisphere 7-day mean 500-mb constant-pressure map for 1-7 December 2011. For reference, approximately one-half of the atmosphere's mass is above and half is below the 500-mb pressure level, and, on a world-wide average, 500 mb pressure is located at a height of 5750 meters above sea level. In this polar projection view, the North Pole is at the center and North America is in the lower portion of the map. The map is a topographical relief map (similar to maps employed in geology to show hills, valleys and steepness of slopes by use of height contour lines) of the 7-day mean height of the 500-mb surface in

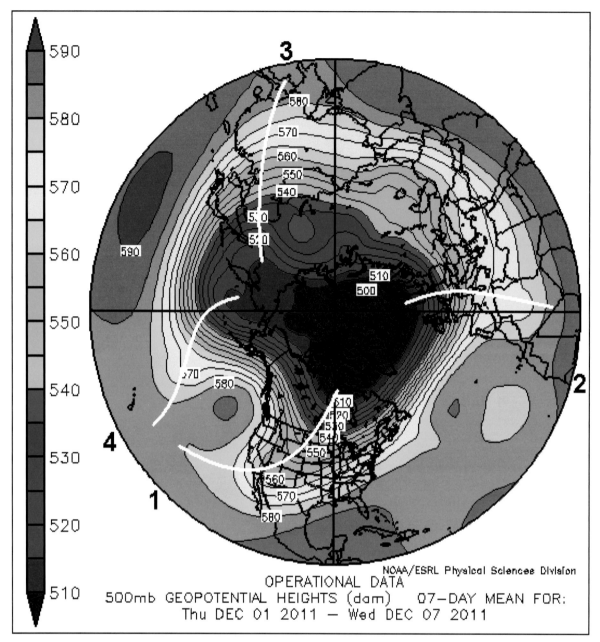

Figure 1.
Northern Hemisphere 7-day mean 500-mb map for 1-7 December 2011. [NOAA]

the atmosphere, depicting an imaginary surface on which the air pressure is everywhere 500 mb.

1. The meandering solid black lines on the Figure 1 map are height contours (lines of constant altitude), and are labeled in tens of meters, that is, the line labeled "570" signifies that everywhere on that line the pressure of 500 mb is located at a height of [(*570*)(*5700*)(*57000*)] meters above sea level.

2. The pattern of contour lines on the map shows that as latitude increases, the height of the 500-mb surface generally [(*increases*)(*decreases*)].

3. It can be assumed that the lower the altitude at which the air pressure is 500 mb, the colder the underlying air. (Cold air is denser than warm air so that air pressure drops more rapidly with altitude in cold air than in warm air.) The Figure 1 contour pattern implies that generally the coldest underlying air is at the [(*lowest*)(*highest*)] latitudes.

On upper-air constant-pressure charts, topographical features can be indentified. **Troughs** are valleys in the constant-pressure surface. **Ridges** are crests or elongated elevations of the constant-pressure surface.

4. Wherever the contour lines form a lobe shape extending towards lower latitudes, the pattern indicates the presence of a topographical [(*ridge*)(*trough*)] at the 500-mb level.

5. Wherever the contour lines curve from lower latitude to higher latitude and back toward lower latitude, the pattern depicts a topographical [(*ridge*)(*trough*)] at the 500-mb level.

The contour patterns on 500-mb maps showing troughs and ridges reveal Rossby waves, each with a wavelength composed of a ridge and trough. Typically, there are from two to five wavelengths in the flow pattern circumscribing Earth at the higher latitudes. In Figure 1, bold and numbered white lines have been drawn on the map to represent the axes of major wave troughs in the higher latitude flow. In this example, there are four Rossby waves whose trough axes (labeled 1, 2, 3, and 4 outside the map border) are shown.

6. At the time of the Figure 1 map, [(*1*)(*2*)(*4*)] Rossby wave(s) stretched over the coterminous U.S. This number of waves is common for the upper-air circulation pattern across the U.S. "lower 48".

7. At the time of this map, the locations of the contour lines and the trough axis over North America show that the 500-mb pressure surface was at lower altitudes in the [(*western*)(*eastern*)] coterminous U.S.

8. The contour configuration during that time period implies that temperatures in the atmosphere between Earth's surface and the 500-mb pressure level were lower in the [(*western*)(*central*)(*eastern*)] coterminous U.S.

9. The western coastal states were being impacted by a [(*ridge*)(*trough*)] whose axis (not drawn on the map) was positioned over the eastern North Pacific Ocean.

While Figure 1 portrays average upper-air conditions during the 1-7 December 2011 time period, it is a good approximation of tropospheric circulation at 500 mb over the Northern Hemisphere on 5 December 2011 when a frontal storm system sprawled across the eastern half of the U.S. brought heavy rains, particularly to the Ohio River Valley.

Figure 2 is a detailed upper-air 500-mb constant-pressure map for the coterminous U.S. for 00Z (7 pm EST) 5 December 2011.

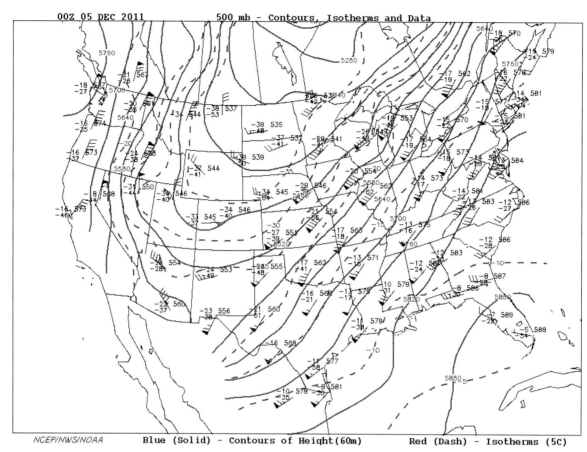

Figure 2.
500-mb constant-pressure chart for 00Z, 5 DEC 2011. [NCEP/NOAA]

10. The Figure 2 map resulted from the analysis of data primarily collected via NOAA's National Weather Service weather balloon radiosonde network. Highlight the 5700-m contour line on the Figure 2 map (one segment labeled along the Illinois-Kentucky border and another segment extending northward from California). Compare Figures 1 and 2, which are for comparable times and, over North America, are based on the same data. The contour-line patterns over the parts of North America appearing on both maps are generally [(*different*)(*similar*)].

11. Arrows with barbs and pennants describe winds reported at 500 mb on the Figure 2 map. The arrows point in the direction towards which the wind is blowing and half-barbs (5 kt), full-barbs (10 kt), and pennants (50 kt) are accumulated on the arrow shaft to denote speed. Decoding wind information on the map shows, for example, winds at 500 mb over central Illinois were blowing at 90 kt from the southwest. The observed winds are generally flowing parallel to the contour lines and strongest where contour lines are [(*farthest apart*)(*closest together*)].

12. Temperatures are reported at each station to the upper left of the arrow point. The 500-mb temperature in the Bismarck, ND area was –39° C while it was –5° C over Miami, FL.

Determine the overall temperature pattern across the U.S. by examining reported station data and the dashed red isotherms. Recalling that relatively low temperatures at the 500-mb level indicate low temperatures in the air between 500 mb and Earth's surface and higher temperatures at 500 mb signify higher temperatures below, it appears that the west-central coterminous U.S. was probably experiencing [(*colder*)(*warmer*)] temperatures than the U.S. east coast states.

13. The location of Rossby waves has a major impact on whether a specific location has relatively warm or cold temperatures and whether storms develop or weaken. Over time, the ridges and troughs of Rossby waves tend to advance eastward. If the Rossby wave examined here moved eastward about a half a wavelength, the surface temperatures in the eastern U.S. would generally [(*increase*)(*remain the same*)(*decrease*)].

Figure 3 is the map of one-day total precipitation ending at 12Z 5 December 2011 from NOAA's Advanced Hydrologic Prediction Service Center for the National Weather Service's Southern Region. The heavy rains brought moderate flooding to portions of eastern Arkansas during the period.

14. The greatest rainfall totals as indicated by the shadings in Figure 3 (moderate red) across east-central Arkansas were in the [(*0.1-0.5*)(*0.5-1.0*)(*2.0-3.0*)(*4.0-5.0*)]-inch range. This amount fell in twenty-four hours with runoff great enough to produce stream and river flooding.

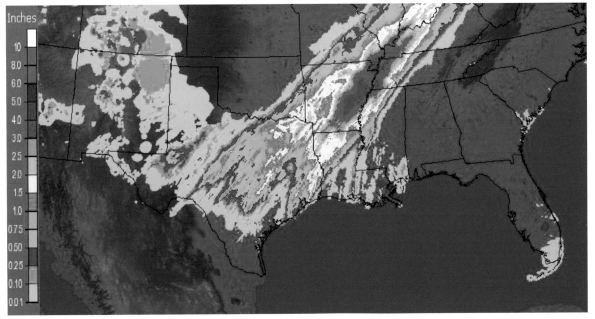

Figure 3.
One-day total precipitation ending at 12Z 5 December 2011 from NOAA's Advanced Hydrologic Prediction Service for NWS' Southern Region.

The pattern of rainfall suggests the general southwest-to-northeast progression of thunderstorms that flowed with the 500-mb winds as shown in Figure 2 over the south-central U.S. These storms tracked along a surface cold front bringing repeated rain episodes to the area. Strong upper air troughs as shown at 500 mb are often associated with outbreaks of storms and severe weather to the east of the axis of the trough.

Rossby Waves and Climate: As can be inferred from the 5 December 2011 precipitation episode, weather in the middle latitudes is strongly influenced by the number, length, amplitude, and location of overlying Rossby waves. Winds in Rossby waves steer cyclones and advect air masses. By the same token, prevailing (or dominating) Rossby wave patterns strongly influence large-scale climate. The basic characteristics of Rossby waves vary considerably with season. Rossby wave ridges and troughs have preferred locations because of the influence of the underlying Earth's surface. These include major mountain ranges (e.g., Tibetan Plateau and Rocky Mountains) and heat sources (e.g., ocean currents in winter and landmasses in summer).

Optional: For additional studies of Rossby waves, the Geophysical Fluid Dynamics Laboratory (GFDL), School of Oceanography, University of Washington, Seattle, WA, provides theoretical foundations and animations showing demonstrations of Rossby waves generated in rotating basins. **Figure 4** is one example of Rossby waves generated in the laboratory.

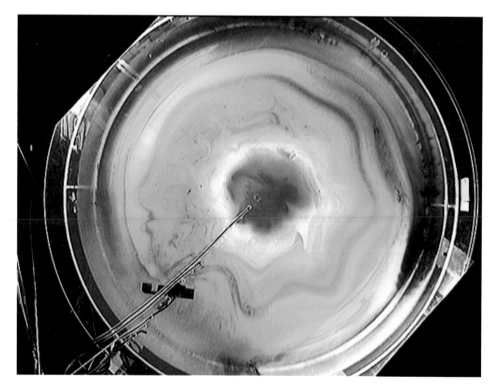

Figure 4.
Rossby waves produced at the Geophysical Fluid Dynamics Laboratory, School of Oceanography, University of Washington.

To view animations, go to: *http://www.ocean.washington.edu/research/gfd/rossby3.html*. There, scroll down to "Waves in fluid nearly at rest" and, in the fourth paragraph, click on "here" in the first sentence to view a 2 Mb animation of a fully developed Rossby wave field produced in the laboratory.

To learn even more about Rossby waves, explore the GFDL website.

Summary: Earth's atmosphere is set into motion primarily through differential heating by the Sun. Part of the motion is the meandering long-wave flow in the middle and upper reaches of the troposphere over the middle and higher latitudes. The flow stretches around the planet and alternately turns left and right as seen from above. This upper-air planetary-scale flow pattern known as Rossby waves, in combination with the near-surface weather systems, provide the transport mechanisms that move heat energy poleward as part of the atmosphere's general circulation. Rossby waves give substance to the concept of the general circulation as a planetary phenomenon, the behavior of which is governed by broad-scale boundary conditions rather than by the local variations of weather.

Another Option: Go to: *http://www.esrl.noaa.gov/psd/map/images/fnl/500z_07.fnl.anim.html*. There, you will see the latest weekly animation of the 500 mb mean heights.

Investigation 7A:

7A - 1

SYNOPTIC-SCALE ATMOSPHERIC CIRCULATION

– HIGH AND LOW PRESSURE SYSTEMS

Driving Question: *What are the synoptic-scale features of climate? Where do these features form, what are their climate-impacting features, and where and when do they prevail?*

Educational Outcomes: To describe the origins and characteristics of the synoptic-scale high and low atmospheric pressure systems. To explain their impacts on local and regional weather and climate, and how these impacts vary over the year.

Objectives: After completing this investigation, you should be able to:

- Identify the synoptic-scale high and low pressure systems that play major roles in determining local and regional climates of middle and high latitudes.
- Describe short-term and seasonal changes in the weather patterns which imply boundary conditions of weather and climate at different times of the year.

Synoptic-Scale Atmospheric Circulation

The atmospheric synoptic scale covers the range between planetary-scale (thousands of kilometers) and mesoscale (tens of kilometers). The features commonly appearing on national weather maps, including highs, lows and fronts, are synoptic in scale.

Weather and climate result from a combination of factors including (a) the directional receipt of solar energy, (b) Earth's rotation and revolution, (c) character of Earth's surface (e.g., topographical relief, land/water surfaces), and (d) atmospheric composition (e.g., clouds, greenhouse gases, aerosols).

The interplay of these factors produces the boundary conditions of weather and climate. While complex, the impacts of these factors can be detected by surveying the state of Earth's climate system at different times, and by observing changes over time.

A. January Snapshot of Weather (and Climate) in the Northern Hemisphere Middle Latitudes:

For a view of typical winter weather, **Figure 1** displays general weather features for the coterminous U.S. and parts of Canada and Mexico at 7:00 a.m. EST on 12 January 2012.

1. The major features of surface weather maps are revealed by analysis of air pressure readings adjusted to the same altitude (i.e., "corrected" to sea level). The drawing of isobars (lines of constant pressure) show pressure patterns over the region covered by the map. These patterns provide information on horizontal forces (called pressure

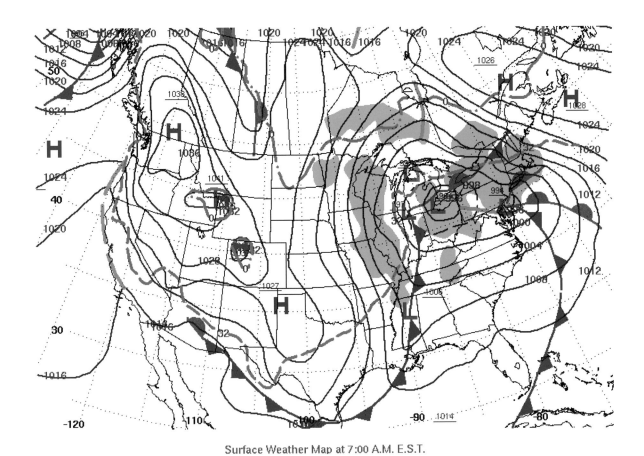

Figure 1.
U.S. Daily Weather Map for 7:00 a.m. EST, 12 January 2012. [HPC/NOAA]

gradient forces) which act to put air into motion. In some places and at some times, isobars completely or partially surround areas of relatively high- and low-pressure and are marked on the map with **H** and **L** labels, respectively. On this Figure 1 map, an expansive and elongated high-pressure area (with highest pressures encircled by 1036-mb isobars) stretched across much of the mountain west. A printed **L** appearing over western Lake Erie identified a broad low-pressure area surrounded by a 996-mb isobar. Other prominent **L**'s were plotted over lower Lake Michigan and northeastern Maryland. The isobar interval (difference in pressure values between neighboring isobars) on the map is [(*4*)(*8*)(*10*)] mb. This is the isobar interval used for pressure analyses on most U.S. surface weather maps.

2. Green shading on the map delineates areas of precipitation. These precipitation areas are generally more closely associated with [(*high-pressure*)(*low-pressure*)] systems.

3. The blue dashed and dot/dashed lines on the map are the 32°F and 0°F isotherms, respectively. The isotherm pattern shows that much of the western half of the coterminous U.S. was subjected to freezing temperatures at map time, with sub-zero cold in parts of Idaho, Wyoming, Colorado, and [(*Michigan*)(*Maine*)(*North Dakota*)].

To retrieve more weather details for this same day, 12 January 2012, go to NOAA's Hydrometeorological Prediction Center's (HPC) Daily Weather Map website: *http://www.hpc.ncep.noaa.gov/dailywxmap/index.html*.

To the left under "Select Date," select "January," "12," and "2012," click on the lower "Get Map" button. The large map that appears is the same as Figure 1.

4. Click on the large map. The new map that appears is for the same time as Figure 1 and has detailed observational data plotted at individual locations. The plotted wind arrows at individual stations (each pointing in the direction towards which air was flowing) show that to the north of the dashed blue 32° isotherm line, across the central Plains states (including Kansas and Nebraska), the winds are generally blowing from the [(*northwest*)(*northeast*)(*southeast*)]. This flow of cold air is an extension of the normal clockwise and outward spiral circulation (as seen from above) generally observed in high-pressure systems, and especially evident on this map in the expansive **H** centered in the northwestern states.

5. Extending out from the low-pressure centers are cold (blue), stationary (alternating red and blue) and warm (red) fronts. Fronts signify boundaries between neighboring air masses with different temperature and/or humidity characteristics. Movement of fronts is indicated by symbols and their positioning on the heavy lines showing the location of fronts. Blue triangles identify a cold front and point in the direction it is moving. Red semi-circles identify a warm front and are on the side of the front towards which it is moving. Alternating red semi-circles and blue triangles pointing in opposite directions label stationary fronts. The wind arrows in the eastern third of the country centered on the lowest-pressure center over western Lake Erie reveal a gigantic overall [(*clockwise*)(*counterclockwise*)] circulation pattern as seen from above. This is typical of atmospheric circulations around centers of surface low pressure in the Northern Hemisphere.

Return to the generalized map by clicking the "Back to Main Page" button above the map on screen or on your browser's back button. Scroll down to the bottom two images.

6. Examine the map to the bottom right showing the total precipitation in the 24-hour period ending on 7 a.m., 12 January 2012. Compare the areas with and without precipitation across the country, especially noting locations of heaviest precipitation. The comparison confirms that the precipitation areas were more closely associated with the [(*high*)(*low*)] pressure areas seen on the 12 January 2012 daily weather map.

Now examine the map at the bottom left showing the upper-air 500-mb constant pressure surface at 7:00 a.m. EST on 12 January 2012. Now click on the 500-mb map for a more detailed view.

7. Comparison of this upper air map with the Figure 1 surface map shows the general relationships between Rossby waves and surface weather features that you were

introduced to in Investigation 6B. Surface storm systems (Lows) tend to be more closely related to the east side of upper air [(*ridges*)(*troughs*)].

Now click on your browser's back button or on "Back to Main Page" above the map to return to the main surface weather map. Starting with the 12 January 2012 map, follow the major weather features seen on the surface map as they progress and evolve over the next day. Do this by clicking on "Next Day" to the upper right. To replay the sequence, first click on the "Previous Day" button to return to the 12 January 2012 map.

8. Note that by the next morning (13 January 2012), the major low-pressure area centered over western Lake Erie moved [(*westward*)(*eastward*)(*southward*)]. In general terms, this is a common direction of movement for weather systems impacting the coterminous U.S.

B. July Snapshot of Weather (and Climate) in the Northern Hemisphere Middle Latitudes:

Now set the HPC Daily Weather Map website to 14 July 2011 for a view of summer weather.

9. Compare the pressure pattern on the July map with that on the 12 January 2012 winter map in Figure 1. The range of pressures and numbers of isobars on the July map are [(*less than*)(*about the same as*)(*more than*)] those on the January map.

10. The closer the spacing between isobars on the maps, the greater the change in air pressure horizontally per unit distance and the greater the horizontal force (called the pressure gradient force) acting on air. This, in turn, results in higher wind speeds. Consequently, it is likely that wind speeds over the map area in January are [(*weaker than*) (*about the same as*)(*stronger than*)] those in July. [Note: You can check this by looking at the detailed surface maps for the two dates.]

11. Now, compare the contour patterns on the 500-mb maps for the two dates. The closer the spacing between neighboring contour lines, the greater the horizontal pressure gradient forces putting air into motion. Consequently, it is likely that wind speeds on the 500-mb constant pressure surface over the map area in January are [(*weaker than*) (*about the same as*)(*stronger than*)] those in July. [Note: You can check this by looking at the detailed 500-mb maps for the two dates by clicking on the small maps to access detailed views showing wind speeds.]

12. Finally, compare the 24-hour precipitation record ending on 7 a.m. for the two dates by clicking on their respective small color-coded precipitation maps to go to enlarged maps showing precipitation areas and station amounts. Consider the overall patterns of precipitation and locations where amounts of 0.75 in. or greater liquid equivalence were observed. They reflect the common observation that greater precipitation amounts tend to be more widely distributed across the U.S. with [(*winter*)(*summer*)] storms. This arises partially because of different prevailing temperatures and the unique relationship between

temperature and the "capacity" of air to hold water vapor, i.e., the higher the temperature, the higher the water vapor "capacity" of air.

Summary: Snapshots of winter and summer weather as examined in this investigation provide evidence of the dramatic seasonal swings in local and regional weather patterns forming the basis of climate that can be traced back to annual cyclical changes in boundary conditions. Observational data, including those provided to the public via the NOAA/HPC Daily Weather Map series, provide the empirical basis of weather. They also demonstrate the intricate connections confirming that weather and climate result from a combination of factors including (a) the directional receipt of solar energy, (b) Earth's rotation and revolution, (c) character of Earth's surface (e.g., topographical relief, land/water surfaces), and (d) atmospheric composition (e.g., clouds, greenhouse gases).

Investigation 7B: SYNOPTIC-SCALE ATMOSPHERIC CIRCULATION

– WAVE CYCLONES AND STORM TRACKS

Driving Questions: *What are the major storm systems, known as wave or extratropical cyclones, of the middle and high latitudes? Where do they form? What are their climate-impacting features? Where and when do they prevail?*

Educational Outcomes: To describe the origins and characteristics of the synoptic-scale storm systems of the middle and high latitudes, called wave or extratropical cyclones. To explain where they occur and their impacts on local and regional weather and climate.

Objectives: After completing this investigation, you should be able to:

- Describe the synoptic-scale wave or extratropical cyclones that play major roles in determining local and regional climates in the middle and high latitudes.
- Provide an overview of the distribution of wave cyclone tracks.
- Describe short-term and seasonal changes in the atmosphere associated with wave cyclone occurrence which identify boundary conditions of weather and climate at different times of the year.

Introduction: The middle and higher latitudes exhibit considerable variability in weather because of poleward flows of energy as Earth's climate system strives to achieve radiational equilibrium with outer space. As described in earlier investigations, these latitudes are subjected to a combination of thermal and turbulent fluid flow, which result in local weather often alternating between fair and stormy episodes. This investigation examines the storm systems, called **wave** or **extratropical cyclones**, which play major roles in transporting heat energy to higher latitudes.

Air residing over uniform surfaces gradually acquires the physical characteristics of the underlying surfaces; be they cold or warm, dry or humid. As gigantic lobes of the conditioned air, called **air masses**, migrate across Earth's surface, they carry these characteristics with them. Where air masses of different character (which produce different air densities) come in contact, a persistent interface, or transition zone, marking the boundary between them and called a **front**, is formed. Along these fronts, synoptic-scale wave (or extratropical) cyclones can form. These are the migratory storm systems common in the middle and high latitudes.

Wave Cyclones: Wave cyclones form as part of the turbulent atmospheric flow in the middle and high latitudes. They exhibit cyclonic circulations around low-pressure centers, counterclockwise as seen from above in the Northern Hemisphere. They form along the boundaries, or fronts, between neighboring air masses and produce a wavelike deformation of the front.

Climate Studies: Investigations Manual 3rd Edition

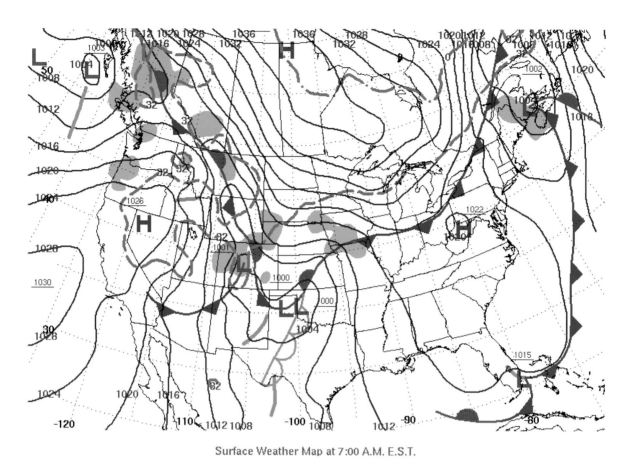

Figure 1.
U.S. Daily Weather Map for 7:00 a.m. EST, 14 April 2011. [NOAA/HPC]

1. **Figure 1** is the NOAA/HPC Daily Weather Map for 12Z (7:00 a.m. EST), Thursday, 14 April 2011. As shown in Figure 1, a frontal system designated by a heavy line stretched its way across the U.S. from north of Maine to Kansas separating a warm air mass to the south and east from a cold air mass to the north and west. A cold front is the leading edge of colder air. The short red portion of the line indicates a warm front across Kansas moving in the direction the symbol is pointing. The air mass boundary is designated a cold front if surface air is flowing so that colder air is replacing warmer air and a warm front if warmer surface air is replacing colder air. The cold front from Maine to Kansas is moving generally towards the [(*northeast*)(*southeast*)].

2. Go to the NOAA/HPC Daily Weather Map website (*http://www.hpc.ncep.noaa.gov/dailywxmap/index.html*). At the left on the webpage, fill in the form for 14 April 2011 and click "Get Map". The map that appears is the same as the Figure 1 map. Now click on the map for a detailed version. Note reported wind directions at stations in the several states surrounding the Low center marked by two Ls in the panhandle of Texas. (For example, at Oklahoma City, OK, the southeast wind was flowing toward the northwest.) The wind directions exhibit a [(*clockwise*)(*counterclockwise*)] circulation as seen from above.

3. A short warm front extending from the Texas low-pressure center indicates relatively warm air was replacing retreating cold surface air in that area. This movement across Earth's surface [(*is*)(*is not*)] consistent with the wind flow described in Item 2.

The orange line with open semicircles extending from west-central Oklahoma across Texas denotes a dry line, a non-frontal boundary separating warm, humid air to the east from warm, but dry air to the west. The dry air is flowing into the system from dry-land areas while the humid air has its origins over the waters of the Gulf of Mexico. This is a frequent feature in the Southwest in summer arising from differences in air density primarily due to variations in water vapor content (the lower the humidity, the greater the density). The resulting density differences often lead to clouds and thunderstorms. Orange dashed lines are extensions of lower pressure showing curvature in isobars, producing low-pressure troughs.

The surface weather map features (low-pressure center, cold and warm fronts, wind pattern) you have examined on the 14 April 2011 map clearly display the initial stage in the life cycle of a **wave cyclone**.

Figure 2 shows the same system 24 hours later. From the NOAA/HPC Daily Weather Map website, fill in the form for 15 April 2011 and get map (or click on "Next Day" above the 14 April map). It is the Daily Weather Map for 7:00 a.m. EST, 15 April 2011, below.

4. Comparisons of the maps in Figure 1 and 2 show that the low-pressure center moved northeastward. The central pressure value is given as a small underlined number near the L. The maps indicate the central pressure of the cyclone decreased [(*4*)(*8*)(*11*)] mb during the 24 hours between map times indicating intensification of the storm system.

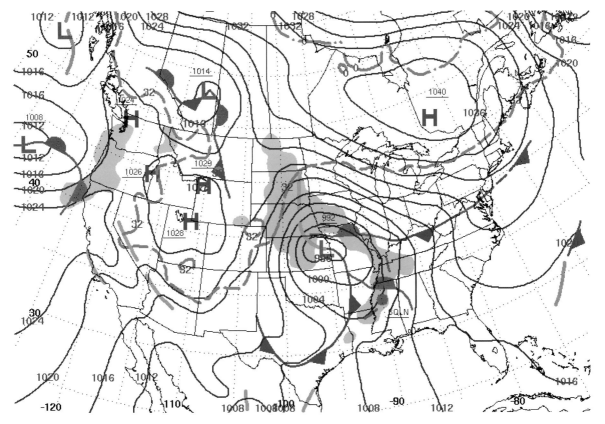

Figure 2.
U.S. Daily Weather Map for 7:00 a.m. EST, 15 April 2011.

5. Click on the 15 April 2011 surface weather map for the detailed version. Wave cyclones typically are divided by the cold and warm fronts into warm and cold sectors. Jackson, in central Mississippi, with a temperature of 72°F, was in the [(*cold*)(*warm*)] sector of this well-developed wave cyclone.

6. Temperature and dew point values reported at individual stations on the two sides of a wave cyclone's cold front often show the contrast between the two air masses in contact. Dewpoint, a humidity measure reported at the 8 o'clock position on map station models, increases as the amount of water vapor in the air increases. The dewpoint at Jackson was 66°F. Compare warm-sector dewpoints with dewpoints behind the cold front, such as at Dallas, TX. They indicate that compared to the warm-sector air mass, the cold air mass was [(*more*)(*less*)] humid. The water vapor being fed into the wave cyclone can be a major energy source that drives the storm system, thereby transporting energy from one place to another across Earth's surface.

7. Note the purple front drawn on the map from the Kansas low-pressure center southeastward. Its alternating triangular and semi-circle symbols face in the direction towards which the front is moving. This is called an **occluded front**, which forms when the faster moving cold front catches up with the slower moving warm front. Its growing length over time is an indication of the eventual maturity and then demise of the wave cyclone. Assuming continued motion in the same northeast direction at about the same speed, it could be predicted that a day later, at 7:00 a.m. EST on 16 April, the low pressure center will be in the general area of [(*Cape Hatteras, NC*)(*the Great Lakes*)].

Figure 3 is a composite of two visible satellite images provided by NCEP/NOAA showing cloud cover at 1815Z (1:15 p.m. EST) on 14 April 2011 (left) and 15 April 2011 (right). The left image is 6 hours after the Figure 1 surface map while the right is 6 hours after the Figure 2 map. An "L" superimposed on each shows the location of the surface low-pressure center at image time.

8. The position of the low-pressure center is marked with an "L" in each view. Note that the cloud patterns suggest the counter-clockwise and inward spiral of air flow surrounding the storm's center. As is typical with wave cyclones, the change in location of cloudiness about the Low center and the position of the center itself indicate movement of the storm system across Earth's surface generally toward the [(*west*)(*east*)].

9. The expansive cloud cover in the center of the U.S., particularly the view on the right, displays a comma (,) shape. This is a common cloud-cover configuration of mature wave cyclones. In the right view, the band of clouds curving from the Low center to the Texas Gulf Coast region would likely align with the [(*cold*)(*warm*)(*occluded*)] front as seen in Figure 2.

Figure 4 shows the same weather system 24 hours after Figure 2. It is the Daily Weather Map for 7:00 a.m. EST, 16 April 2011.

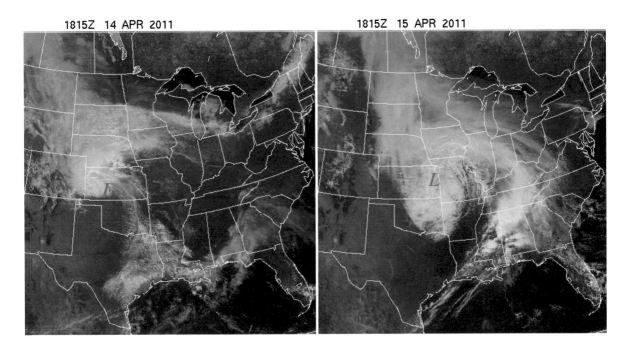

Figure 3.
Composite visible satellite images at 1815Z on 14 April 2011 (left) and 15 April 2011 (right).

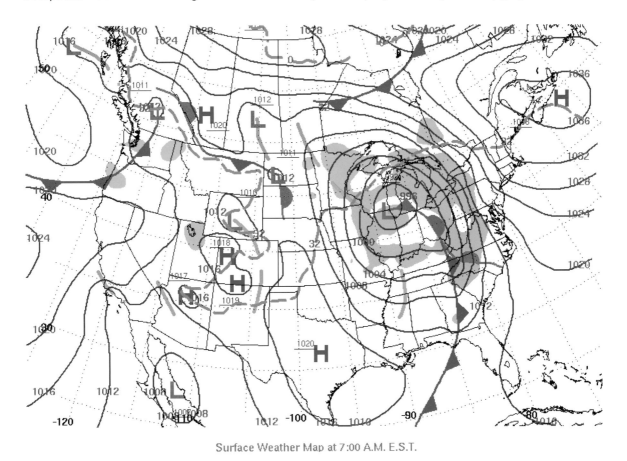

Figure 4.
U.S. Daily Weather Map 7:00 a.m. EST, 16 April 2011.

10. Figure 4 shows the wave cyclone in a late stage of its life cycle. The location of the low-pressure center [(*__did__*)(*__did not__*)] track as expected in item 7.

Figure 5 shows the same weather system 24 hours after Figure 3, at 7:00 a.m. EST, 17 April 2011.

The low-pressure center was located north of the Great Lakes in Canada by 17 April with the occluded front beginning to dissipate (shown by the broken purple line). A new Low center had formed over New England and the system was moving rapidly to the northeast and out to sea. By 18 April, the storm remnants were passing the Canadian Maritime Provinces into the Atlantic.

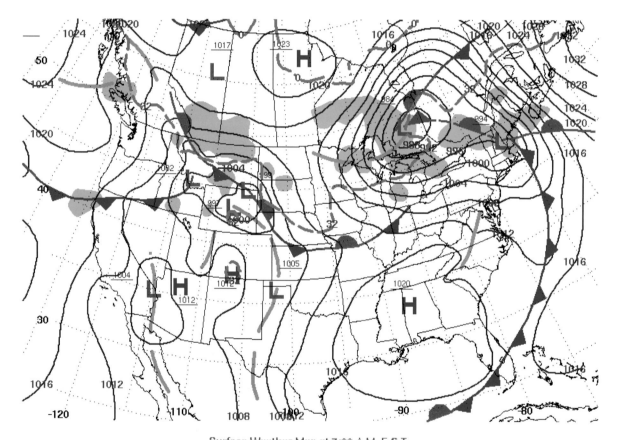

Figure 5.
U.S. Daily Weather Map for 7:00 a.m. EST, 17 April 2011.

11. Figure 5 shows the wave cyclone in a late stage of its life cycle. As is typical of most wave cyclones, its entire lifecycle was completed in a few [(*__hours__*)(*__days__*)(*__weeks__*)].

Extratropical Storm Tracks: As evident in the April 2011 wave cyclone, extratropical storm systems are integral components of the global atmospheric circulation. Although relatively short-lived, these migratory synoptic-scale systems can impact the weather and climate over vast areas. Over one-third of the coterminous U.S. directly experienced some combination of precipitation, temperature change, wind variations, pressure fluctuations, etc., as this April 2011 storm developed and evolved through its life cycle while tracking across Earth's surface.

Where do these storms form? What are their typical paths or tracks? Is their occurrence more frequent in some places than others? Do they play major roles in determining local and regional climate? Examining the tracks of extratropical storms can help us find answers to these questions while contributing to better weather forecasting and climate prediction.

Figure 6 displays imagery from CPC/NOAA in which air pressure analyses are employed to place red dots showing positions where extratropical storms formed (cyclogensis) and blue dots where they lost cyclonic identity (cyclolysis). The maps show cyclogensis and cyclolysis locations during the 10, 30 and 90 days prior to 1 May 2011.

12. The 90-day map essentially pinpoints where storm systems began and ended during the 2011 winter-spring seasonal transition. The distribution of red dots implies that cyclogensis is not a random event, i.e., the pattern of dots shows some clustering that would not be expected if storm development occurred purely by chance. For example, during this 90-day sample it appears that late winter-early spring storms in the U.S. more frequently formed in the [(*western*)(*eastern*)] half of the country.

13. Although not marked on these maps, the storm systems moved so their tracks (paths) typically extended from red dots to blue dots somewhere further east. Now draw a straight line across the 90-day image at 45° N. Note which colored dots (red or blue) dominate south of 45° N and which dominate north of the 45° latitude line worldwide. This pattern implies that the storm systems tended to move toward [(*lower latitudes*)(*due east*)(*higher latitudes*)]. This is typical of extratropical cyclones.

For the latest maps showing cyclogensis and cyclolysis during the past 10, 30, and 90 days, go to: *http://www.cpc.noaa.gov/products/precip/CWlink/stormtracks/combined_cgl.gif*.

Optional: Detailed seasonal and monthly maps of extratropical storm tracks in both the Northern and Southern Hemispheres covering the time period 1961-1998 can be found at: *http://data.giss.nasa.gov/stormtracks/*. We recommend that you visit this NASA site and explore the information provided. Look for overall seasonal patterns in the storm tracks [e.g., latitudinal differences between winter (December, January, February) and summer (June, July, August)].

Summary: Studies of storm tracks show how ubiquitous extratropical cyclones are in the middle and higher latitudes, but they also reveal that considerable variation exists concerning where they form, the paths they take, and where they terminate. These variations imply there are boundary conditions operating, including those related to differences in Earth's surface properties (e.g., ocean, land) and topographical barriers (mountains).

7B - 8

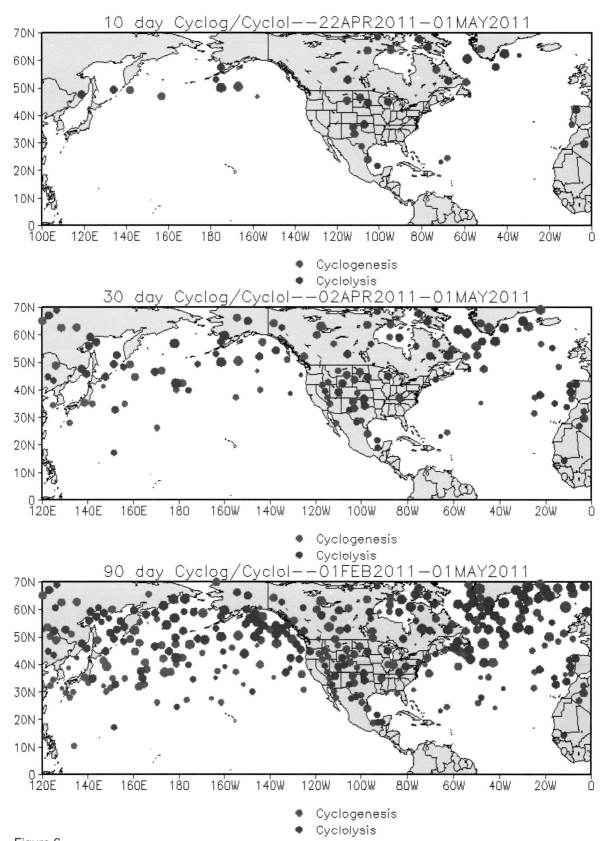

Figure 6.
Maps showing cyclogensis and cyclolysis during three time periods (10, 30 and 90 days) ending on 1 May 2011. [CPC/NOAA]

8A - 1

CLIMATE AND AIR/SEA INTERACTIONS

– INTER-ANNUAL TO DECADAL CLIMATE VARIABILITY

Driving Questions: *How do interactions between the ocean, atmosphere, and Earth's surface produce inter-annual and longer climate variability? What are examples of such short-term climate variability on local, regional, and global scales?*

Educational Outcomes: To describe the origins and characteristics of inter-annual and decadal variability. To explain their impacts on local, regional and global weather and climate.

Objectives: After completing this investigation, you should be able to:

- Describe fundamental concepts concerning climate fluctuations that have been observed lasting beyond a year to decades.
- Explain inter-annual climate variability as exemplified by El Niño – Southern Oscillation (ENSO).
- Describe local, regional, and global impacts or teleconnections of inter-annual climate variability phenomena.

Variability in Earth's Climate System

Earth's climate system shows considerable variability on all time scales. As stated previously, the primary drivers of climate are the direct impacts of (a) incident solar radiation, (b) Earth's rotation and revolution, (c) the character of Earth's surface, and (d) atmospheric composition. These boundary conditions underlie cycles of climate components (e.g., temperature, precipitation) that adhere strictly to diurnal and annual time periods. These essentially determine the multi-year mean state of the climate system (including annual and seasonal mean values of climatic components).

But there are fluctuations that occur in the climate system with time scales lasting greater than a year (inter-annual) to decades and longer. They manifest conditions, or modes, with distinctive life times and spatial patterns. These are departures from the mean state of the climate system that are related to variations in the general circulations of the atmosphere and ocean, and the conditions at Earth's surface such as sea surface temperatures (SST) and snow and ice cover. These modes arise from the forcings and responses, including feedbacks, in the system and the extent to which the slower acting components (e.g., ice, deep ocean, and vegetation) become entrained with the more rapid variations of the atmosphere and ocean surface.

The ocean, because of its attributes, including its great mass, huge heat capacity, fluidity, and thermal inertia, is a crucial participant in producing climate fluctuations that run their course on inter-annual and longer time scales. These fluctuations are significant to societies and ecosystems through their impacts on water resources, food supply, energy demand, human health, biological systems, and national security, among other issues.

El Niño – Southern Oscillation (ENSO): Perhaps the best known and most readily detected inter-annual climate variability phenomenon is **El Niño – Southern Oscillation (ENSO)**, a large-scale, persistent disturbance of ocean and atmosphere in the tropical Pacific Ocean. ENSO is governed by large-scale ocean dynamics and coupled ocean-atmosphere interactions which result in periods of anomalously warm and cold conditions (or phases) in the tropical Pacific Ocean on a quasi-periodic basis. While its different phases have major impacts on the tropical Pacific and bordering land environments, there are also impacts worldwide. Globally, ENSO is the largest single contributor to inter-annual climate variability.

The term *El Niño* originally described a short-term, weak warming of ocean water that ran southward along the coast of Peru and Ecuador around Christmas resulting in poor fishing. Now, the term is incorporated into the more comprehensive general term El Niño - Southern Oscillation (ENSO), which exhibits a warm phase (traditional El Niño), a cold phase (termed La Niña) and a neutral or long-term average phase. The "SO" in ENSO refers to the Southern Oscillation, an interannual see-saw in tropical sea level pressure between the eastern and western portions of the tropical Pacific. In general, the terms El Niño and La Niña may be used for particular phase occurrences.

El Niño modes last an average of 12 to 18 months and occur about once every two to seven years. Ten occurred during a recent 42-year period, with one of the most intense on record in 1997-98. Sometimes El Niño is followed by La Niña, a period of unusually strong trade winds and vigorous upwelling in the eastern tropical Pacific. During La Niña, changes in SST and extremes in weather are typically opposite those observed during El Niño.

A persistent ENSO phase can be accompanied by major shifts in planetary-scale atmospheric and oceanic circulations and weather extremes. Because of the importance of ENSO, the tropical Pacific Ocean is now monitored continuously via about 70 tethered (anchored) buoys in NOAA's Tropical Atmosphere Ocean (TAO) Project. Operational definitions of ENSO phases have been agreed to by the World Meteorological Organization (WMO) Commission for Climatology. They are defined based on three-month averages of SST departures from normal for a critical region (ENSO 3.4 region: 120W-170W, 5 N-5S) of the equatorial Pacific as follows:

El Niño (the warm phase) is characterized by a positive SST departure from normal, equal to or greater than 0.5 Celsius degrees, averaged over three consecutive months.

La Niña (the cold phase) is characterized by a negative SST departure from normal, equal to or greater than 0.5 Celsius degrees, averaged over three consecutive months.

Figure 1 displays the average SST and wind patterns for December 2011 based on observations from NOAA's TAO/Triton equatorial Pacific buoy array.

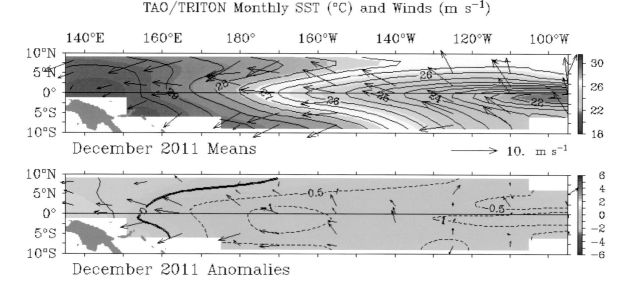

Figure 1.
Tropical Pacific monthly SST and winds during December 2011.

1. The upper panel in Figure 1 depicts SST and winds across the tropical Pacific, and shows the temperature patterns as averaged over the month of December 2011. Isotherms are drawn at a 0.5 C° interval to show the temperature pattern. Temperatures range from above 29.5 °C in the far western Pacific to less than [(*25.5*)(*24.0*)(*22.0*)] °C along the equator near 100 deg W.

2. Average wind directions are indicated by arrows with their lengths proportional to the wind speeds. The prevailing surface winds across most of the tropical Pacific during December 2011 were towards the [(*east*)(*west*)].

The lower panel in Figure 1 denotes anomalies. An **anomaly** is a departure from the long-term average condition. Anomaly patterns are depicted with lines also drawn with half-degree intervals. Positive (warmer than normal) anomalies are analyzed with solid lines and negative (colder than normal) anomalies with dashed lines. The heavy solid black line, labeled "0", denotes locations where SST values were the same as the long-term averages.

3. With a pencil and a straight edge, mark off the ENSO 3.4 region (120W-170W, 5 N-5S). The average anomaly in the region is [(*warmer than +0.5*)(*between 0 and –0.5*) (*colder than –0.5*)] Celsius degrees.

4. Anomalies in the ENSO 3.4 region similar to those in December 2011 had persisted for at least three months. Based on the definitions stated earlier, [(*El Niño*)(*La Niña*)] conditions existed in the tropical Pacific in December 2011. [Note: Tropical Pacific conditions made a transition towards ENSO-neutral during April 2012.]

Figure 2 and **Figure 3** are presented to demonstrate especially strong El Niño and La Niña episodes, respectively, from recent years.

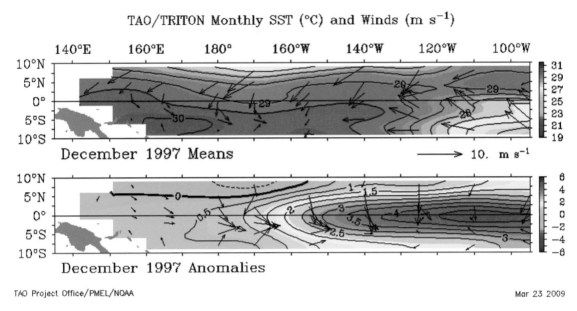

Figure 2.
Monthly SST and winds during a significant El Niño (warm phase) episode.

5. Figure 2 shows tropical Pacific conditions midway through a strong El Niño (warm phase) episode in 1997-98. The upper panel shows warm temperatures generally across the Pacific, with the lower panel showing anomalies in the ENSO 3.4 region which are [(*warmer than +0.5*)(*between 0 and –0.5*)(*colder than –0.5*)] Celsius degrees.

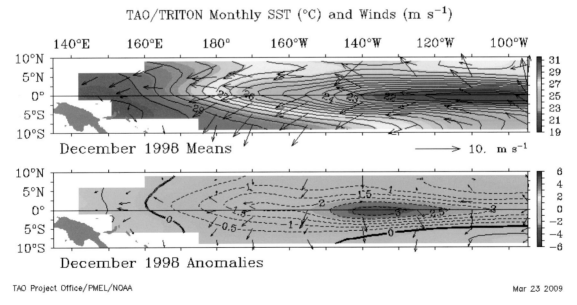

Figure 3.
Monthly SST and winds during significant La Niña (cold phase) episode.

6. Figure 3 shows tropical Pacific conditions midway through a strong La Niña (cold phase) episode during 1998-99. The upper panel shows warm temperatures in the western Pacific and colder temperatures along the equator to the east. The lower panel shows anomalies in the ENSO 3.4 region which are [(*warmer than +0.5*)(*between 0 and –0.5*) (*colder than –0.5*)] Celsius degrees.

7. Compare the wind arrows in the upper panels of Figures 2 and 3. They show that during [(*El Niño*)(*La Niña*)] episodes the wind blows more strongly towards the west across all or most of the tropical Pacific.

For the most recent monthly SST and Winds TAO depiction, go to: *http://www.pmel.noaa.gov/tao/jsdisplay/*. There, click on "Lat Lon plots". On the new page, click on "Monthly" and then, "Make plot!". Click on image to enlarge it. You can also get the latest 5-day average depiction at the same website.

Teleconnections: **Teleconnection** is the name given to statistically significant correlations between weather events that occur at different places around the globe. There are a number of such correlations that appear during ENSO episodes. They result from the changing position of the major heat source (identified by high SST) in the tropical Pacific. These interactions between ocean and atmosphere have a ripple effect on climatic conditions in far flung regions. Shifts in tropical rainfall affect wind patterns over much of the globe. Waves in the air flow pattern determine the positions of the monsoons, storms tracks, and belts of strong upper-air winds (jet streams) which overlie borders between warm and cold air masses at Earth's surface.

El Niño (warm phase) and La Niña (cold phase) teleconnections are summarized in **Figure 4** and **Figure 5**, respectively.

8. According to Figures 4 and 5, ENSO-related teleconnections in North America are more extensive during the Northern Hemisphere [(*summer*)(*winter*)].

9. El Niño (warm phase) related teleconnections are typically stronger, more frequent, and longer lasting than those during La Niña (cold phase) episodes because El Niño events themselves are typically stronger, more frequent, and longer lasting. The teleconnections are also different. If the U.S. southeastern states were experiencing drought, the onset of a Northern Hemisphere winter-season [(*El Niño*)(*La Niña*)] might be welcomed.

10. One part of the coterminous U.S. which appears least likely to be impacted by ENSO events is the [(*northwestern*)(*southwestern*)(*central*)] states.

11. One country that is most impacted by teleconnections to some extent by all El Niño (warm phase) and La Niña (cold phase) episodes is [(*Brazil*)(*Indonesia*)(*Japan*)].

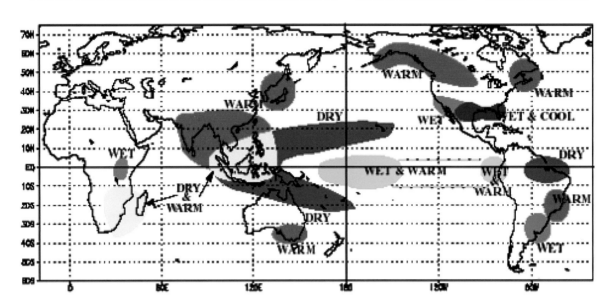

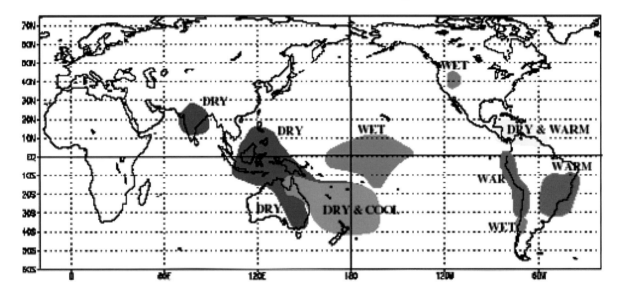

Figure 4.
December-February and June-August El Niño (warm phase) typical teleconnections.

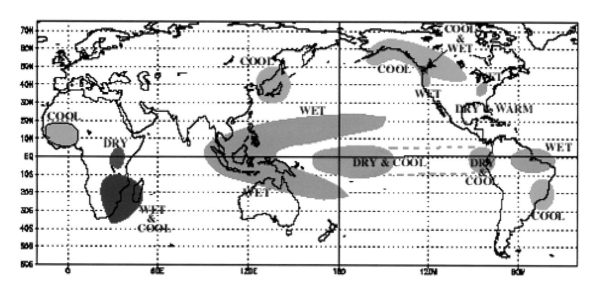

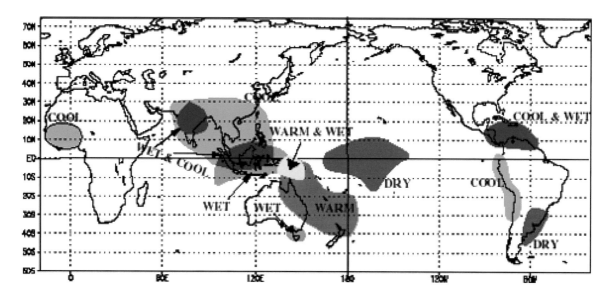

Figure 5.
December-February and June-August La Niña (cold phase) typical teleconnections.

ENSO Impacts on the U.S.: Detailed statistical analyses have been made by NOAA's Climate Prediction Center to describe strong ENSO temperature and precipitation teleconnections. These are available in map forms at: *http://www.cpc.ncep.noaa.gov/products/predictions/threats2/enso/elnino/index.shtml*. As an optional activity, we suggest that you examine maps to determine what, if any, ENSO-related teleconnections have been detected in your geographical area.

Decadal Variability: Additional variabilities in the climate system in certain regions over time frames of several decades have been identified. These periodicities are just beginning to be documented and understood. Regional differentials in sea-level air pressure and/or sea surface temperatures have been associated with upper atmospheric wind patterns or surface ocean conditions in the North Atlantic, North Pacific, Arctic and Antarctic areas.

Summary: Fluctuations occur in the climate system with time scales on the order of more than a year (inter-annual) to decades and longer. Ocean-atmosphere interactions play major roles in causing these climate modes with distinctive life times and spatial patterns. The El Niño – Southern Oscillation is the most significant inter-annual climate variability phenomenon. While the climates of the locations where these fluctuations take place are directly impacted, they can also impact weather and climate in far flung places. These distant impacts are referred to as teleconnections.

These fluctuations are significant to societies and ecosystems through impacts on water resources, food supply, energy demand, human health, biological systems, and national security.

Investigation 8B:

8B - 1

COASTAL UPWELLING AND COASTAL CLIMATES

Driving Questions: *How do winds produce upwelling and downwelling in the ocean? What is the significance of upwelling for marine productivity? What are possible impacts on local climate?*

Educational Outcomes: To describe upwelling and downwelling in coastal areas. To describe relationships between upwelling and marine productivity. To describe the impacts of upwelling and downwelling on California coastal climates.

Objectives: After completing this investigation, you should be able to:

- Demonstrate the causes of coastal upwelling and downwelling.
- Describe the influence of the prevailing wind and Coriolis Effect on upwelling and downwelling.
- Describe the impacts of upwelling and downwelling on coastal climates.

Introduction: In some near-shore areas of the ocean, coastal orientation, prevailing wind, and Earth's rotation combine to influence vertical ocean circulation. In these regions, the wind sometimes transports water in the upper 10 to 100 m (33 to 330 ft) away from the coast, to be replaced by cooler water from below. This process, called **coastal upwelling**, brings to the sunlit surface nutrient-rich water, spurring biological productivity. At other times and places, the wind transports near-surface water towards the coast, causing warm surface waters to pile up and sink. This process, known as **coastal downwelling**, thickens the layer of nutrient-deficient water, reducing biological productivity.

Upwelling and downwelling can be accompanied by dramatic changes in the local weather and climate. Upwelling and downwelling are also associated with broad-scale atmosphere/ocean interactions (e.g., El Niño and La Niño) that have regional and even global impacts on precipitation patterns and other components of the water cycle.

In this investigation, we examine coastal upwelling and downwelling by looking at the contribution of coastline orientation, prevailing winds, and Earth rotation. We also describe the impacts on California coastal areas.

Materials: Scissors and paper brad or a paper clip with its inside end bent at a right angle to the rest of the clip.

Figure 1 (last page of this Investigation) provides a **Model Ocean Basin** manipulative. Use scissors to separate the top and bottom diagrams along the dashed line and to remove the "cut out" areas in the top diagram along the dashed lines. The top block diagram represents the ocean surface with a vertical cross-section through a model ocean basin.

8B - 2

1. According to the cardinal direction arrows in the upper left hand corner of the top block diagram, the east boundary of any ocean basin is the land's [(*eastern*)(*western*)] coast.

Use a pencil point to poke a small hole through the centers (each marked with a ⊕) of the two diagrams. Lay the top diagram (Model Ocean Basin) directly over the bottom diagram (arrows) so the center points of the two coincide. To hold the two together and to provide an axis for rotation, place a paper brad pin through the holes you made on the diagrams. [If you are using a bent paper clip, lay the device flat on a table or desk so the clip doesn't fall out.]

2. While holding the top diagram stationary, rotate the bottom diagram through all 4 positions. Note that there are two different hemispheres, North (*N*) and South (*S*), each with two different wind directions and two different coasts possible. Hence, the total number of different combinations of hemispheres, wind directions, and coasts possible in our model is [(*2*)(*6*)(*8*)].

3. Everywhere except at the equator, surface ocean water set in motion by the wind will be deflected by Earth's rotation. This deflection is called the **Coriolis Effect**. Turn the bottom diagram until a Northern Hemisphere combination appears, that is, when "N" appears in the upper right window. Compare the wind direction and the direction of the near-surface flow of water. (This is best seen by imagining yourself on the tail of the wind arrow and facing forward.) The near-surface flow of water is about 90 degrees to the [(*right*)(*left*)] of the wind direction.

4. Predict the flow direction of near-surface water produced by wind blowing from the opposite direction in the same hemisphere. Your prediction is that the near-surface flow will be about 90 degrees to the [(*right*)(*left*)] of the wind direction.

5. To check your prediction, rotate the bottom diagram until the other "N" appears in the window. From what you have learned so far in this investigation, wind-driven flow of near-surface water in the Northern Hemisphere is about 90 degrees to the [(*right*)(*left*)] of the wind direction.

6. Repeat the last three steps for the Southern Hemisphere. Again predict and note the difference between the wind direction and the direction of flow of near-surface water. The wind-driven flow of near-surface water in the Southern Hemisphere is about 90 degrees to the [(*right*)(*left*)] of the wind direction.

7. Near-surface water transported away from the coast tends to be replaced by cooler water from below in a process called upwelling. Rotate the bottom diagram to a position showing the wind blowing from south to north in the Southern Hemisphere. This combination will produce upwelling along the land's [(*east*)(*west*)] coast.

8. Coastal upwelling of nutrient-rich water stimulates the growth of marine plants which support fisheries. Now rotate the underlay to determine that in the Northern Hemisphere on the west coast of Africa, upwelling and increased productivity will be generated by a wind blowing from [(*south to north*)(*north to south*)].

9. When the wind transports near-surface water towards a coast, the warm surface layer thickens, decreasing biological productivity. This process is called downwelling. Rotate the underlay to a position showing the wind blowing from south to north in the Northern Hemisphere. This combination will produce downwelling along the land's [(*eastern*)(*western*)] coast.

10. Along the coast of central and northern California, surface winds generally blow from north to south in the summer and from south to north in the winter. The season of warm water transport towards the California coast and downwelling for this region is [(*summer*)(*winter*)].

11. The season when cold coastal upwelling water along the California coast cools the overlying air to saturation and produces frequent fog is [(*summer*)(*winter*)].

12. The Southern Hemisphere's trade winds, between the equator and 30 degrees S, have a strong component blowing from south to north. This causes upwelling, high biological productivity, and abundant fish harvests along the [(*western*)(*eastern*)] coasts of Africa and South America.

Coastal Upwelling from a Different Perspective: As part of its NASA-sponsored **Satellite Observations in Science Education**, the University of Wisconsin-Madison Cooperative Institute for Meteorological Satellite Studies (CIMSS) leads an effort to improve the teaching and learning of the Earth system through quality educational resources that make use of satellite observations.

To learn more about coastal upwelling via a learning activity developed in the Satellite Observations in Science Education Program, go to: *http://www.ssec.wisc.edu/sose*. Under the "Learning Activities" heading, scroll down to and click on "5) Coastal Upwelling". Proceed as directed through Module A entitled, "The Biological Pump and the Significance of Upwelling".

Optional: After completing this CIMSS module, consider clicking "Main Menu" on the screen at the lower left, then on "5) Coastal Upwelling", and complete Modules B and C.

Summary: Coastal upwelling and downwelling of ocean water have significant effects on coastal weather and climates.

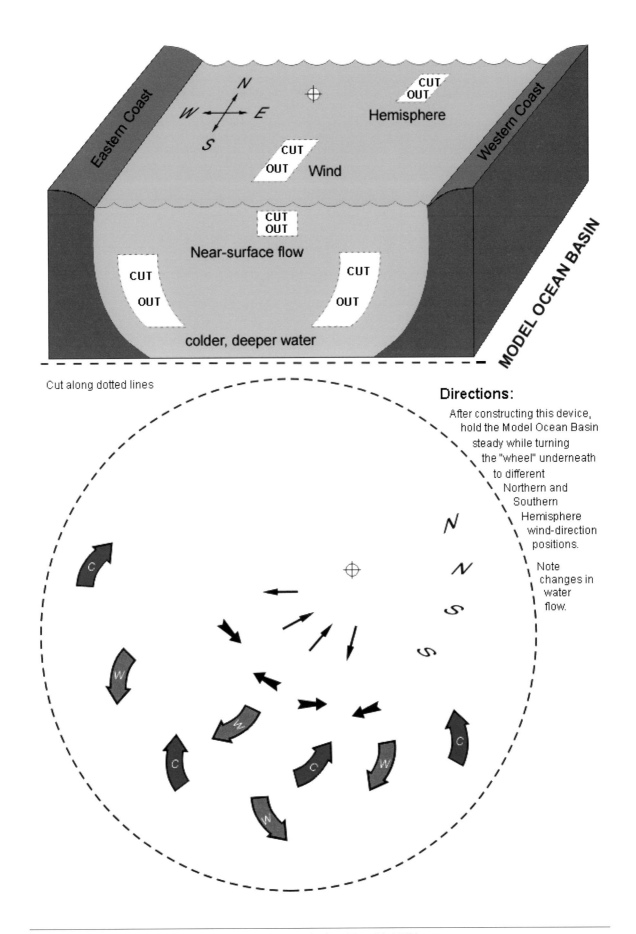

9A - 1

Investigation 9A: PETM: A POSSIBLE ANALOG TO MODERN CLIMATE CHANGE

Driving Questions: *What is the paleoclimatic event called PETM (Paleocene-Eocene Thermal Maximum) and how was it discovered via analysis of deep sea sediment cores? Are there similarities between PETM and current anthropogenic loading of carbon dioxide in the atmosphere and ocean?*

Educational Outcomes: To describe the PETM and how it was determined from the geologic records archived in deep-sea sediment cores. To examine any similarities between what happened during the PETM and the current upward trends in atmospheric and oceanic carbon dioxide.

Objectives: After completing this investigation, you should be able to:

- Describe ways in which analyses of deep-sea sediment cores are employed in reconstructing past climates, using PETM as an example.
- Compare possible similarities between PETM and modern climate change.

Reconstructing Past Climates

The methods and findings of paleoclimatological studies are providing valuable perspectives on the present climate and climate change. Deep-sea sediment cores (sample shown in **Figure 1**) systematically acquired since the late 1960s through the Integrated Ocean Drilling Program and its predecessors, provide proxy (e.g., based on indirect evidence) climate data going back tens of thousands to hundreds of millions of years. Stable isotopes (*isotopes* are atoms of the same element with different masses because of differences in the number of neutrons in their nuclei) of oxygen and carbon are commonly used to acquire proxy climate records in paleoclimatology.

Figure 1.
Deep sea sediment core showing the onset of the PETM as an abrupt change from light to dark sediment. [American Museum of Natural History]

Of particular interest has been the comparison between temperatures and carbon dioxide concentrations in the reconstruction of past climates. The main reason is that some past climatic episodes exhibit similarities to the modern-day rapid increases in the flux of carbon into our atmosphere and ocean. The **Paleocene-Eocene Thermal Maximum (PETM)**, occurring early in the Cenozoic Era (65.5 million years ago to the present) near the transition from the Paleocene to the Eocene Epoch, was a geologically brief period of widespread and extreme climatic warming that was associated with massive atmospheric greenhouse gas input into Earth's climate system. The estimated magnitudes of these carbon releases are comparable to releases due to human activity at the present time and expected this century. The PETM serves as a reference showing the effect of a rapid increase of fossil carbon into the atmosphere, although it is dwarfed by comparison to the present rate of CO_2 releases into Earth's atmosphere. The onset of PETM was marked with temperatures increasing by several Celsius degrees, at least 5 C° (9 F°) during a time period of between 1000 and 10,000 years. The mass of increased carbon was sufficiently large to lower the pH of ocean waters, driving widespread dissolution of seafloor carbonates. The geologic record also shows that between 100,000 and 200,000 years passed for the ocean to fully recover from the PETM through natural carbon sequestration (removal) processes. That time scale is similar to a forecast of the time which will be required for the modern-day ocean to return to its pre-industrial age condition. [Intergovernmental Panel on Climate Change (IPCC), *Climate Change 2007: The Physical Science Basis Report*, pp. 442-443, and the National Academies' *Understanding Earth's Deep Past: Lessons for Our Climate Future* (2011), pp, 69-71]

1. **Figure 2** depicts the PETM as a temperature spike of several Celsius degrees that took place approximately **[(*40*)(*50*)(*56*)]** million years before present. [Note that in Figure 2 the most ancient time is to the right and the most recent is to the left, that is, the direction of time is from right to left.]

Much of what is known about the PETM is from the study of the stable isotopes of oxygen and carbon obtained from deep-sea sediment cores. From cores that include fossils composed of calcite ($CaCO_3$), determinations of the proportions or ratios of carbon isotopes ($^{13}C:^{12}C$) or oxygen isotopes ($^{18}O:^{16}O$) can be made. The ratios of these isotopes vary depending on physical conditions and species involved. Because organisms build their shells in isotopic equilibrium with the seawater in which they live, they reliably record history that can be read to reconstruct past climates.

For a detailed overview of the employment of stable isotopic analysis in paleoceanography, go to the Consortium for Ocean Leadership's *Abrupt Events of the Past 70 Million Years* at http://www.oceanleadership.org/education/deep-earth-academy/educators/classroom-activities/undergraduate/abrupt-events-of-the-past-70-million-years-evidence-from-scientific-ocean-drilling/. There, click on "Download Abrupt Events".

2. Page 2 of this Deep Earth Academy's overview describes how evaporation of seawater and build-up of land ice during glacial climatic episodes results in seawater that is enriched in **[(^{18}O)(^{16}O)]**. Therefore, ocean microfossils that lived during those glacial

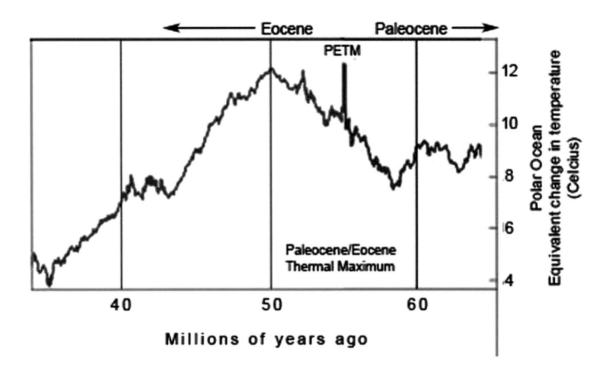

Figure 2.
Temperature change curve showing PETM spike. [Mongabay.com. Rhett A. Butler - San Francisco, CA. 1999-2008]

episodes have calcite shells with relatively higher proportions of ^{18}O compared with those ocean microfossils that lived during interglacial climatic episodes.

3. From the discussion in item 2 above, it can be inferred that the oxygen isotope ratio (^{18}O:^{16}O) is a proxy for [(*temperature*)(*glacial ice volume*)(*both of these*)].

Figure 3, reported in the IPCC *Climate Change 2007, The Physical Science Basis*, p. 443, shows detected changes in carbon and oxygen isotope ratios, and percentage of calcium carbonate saturation, before, during, and after the PETM event. Note the time scale in this Figure 3, in megannum (Ma), a unit of time equal to one million (10^6) years, with the most ancient to the left and most recent to the right. The direction of time is from left to right.

In Figure 3, stable isotopes of oxygen and carbon in samples were measured by comparing the ratios of ^{13}C:^{12}C and ^{18}O:^{16}O relative to those in a known standard. These "del" values (δ) are reported in units of parts per thousand (‰).

4. Examine the top graph in Figure 3. Isotopic studies of bottom-dwelling foraminifera fossils recovered from deep-sea sediment cores from sites described in the Figure 3 caption show that with the onset of PETM there was a rapid [(*increase*)(*decrease*)] in the carbon isotope δ (del) value (note scale on left increases upward). This is indicative of a large increase in the concentration of the atmospheric greenhouse gases carbon dioxide (CO_2) and methane (CH_4).

5. The middle panel has scales for two different measures. Note that the oxygen isotope δ (del) scale on the left increases downward, while the water temperature scale on the right increases upward. The middle panel shows that a relatively rapid decrease in oxygen isotope δ values was associated with a dramatic [(*increase*)(*decrease*)] in seawater temperature of approximately 5 Celsius degrees.

6. The bottom panel in Figure 3 reports the percentage of calcium carbonate ($CaCO_3$) in solution in seawater compared to the amount if the liquid were saturated with $CaCO_3$. The third panel shows that during the period prior to and the period after the PETM event, ocean waters at the two depths measured (4.8 km and 2.6 km) were approximately [(*0%*)(*50%*)(*90%*)] saturated with calcium carbonate ($CaCO_3$).

7. Under the full impact of PETM, the ocean waters at those depths were approximately [(*0%*)(*50%*)(*90%*)] saturated with $CaCO_3$. This condition, brought on by the ocean's absorption of carbon dioxide, lowered seawater pH (increased acidity) and caused widespread dissolution of seafloor carbonates.

How did the PETM happen? The IPCC states that, "Although there is still too much uncertainty in the data to derive a quantitative estimate of climate sensitivity from the PETM, the event is 'a striking example of massive carbon release and related extreme warming.'" [*Climate Change 2007, The Physical Science Basis*, p. 442]

Summary: The similarities between the PETM and the comparable current increases in atmospheric and oceanic carbon makes the PETM an analog worthy of study as we face climate change.

The possible sources of such a large mass of carbon could be from the release of methane (CH_4) from decomposition of hydrate deposits in seafloor sediments, CO_2 from volcanic activity, or oxidation of organic-rich sediments. In **Part B** of this investigation, we will explore gas hydrates as components of Earth's climate system.

Optional: To view an approximately 8-minute video entitled ***PETM: Unearthing Ancient Climate Change***, go to *http://www.amnh.org/sciencebulletins/?sid=e.f.PETM.20081126&src=1*. This American Museum of Natural History Science Bulletin presentation describes the work of paleontologists, paleobotanists, soil scientists and other researchers in Wyoming's Bighorn Basin as they reconstruct how climate, plants, and animals changed during the PETM.

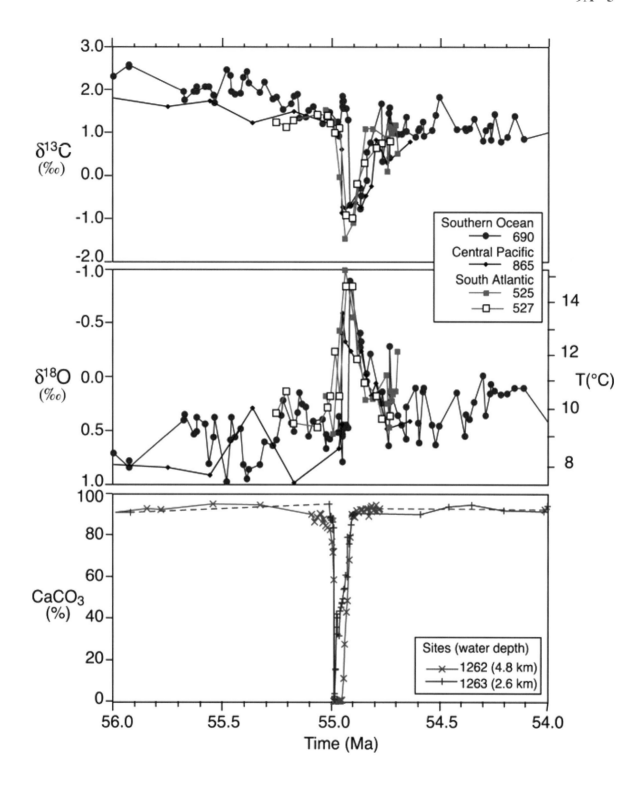

Figure 3.
The PETM as recorded in benthic (bottom dwelling) foraminifera (*Nuttallides truempyi*) isotopic records from sites in the Antarctic, south Atlantic and Pacific. [IPCC, *http://www.ipcc.ch/graphics/ar4-wg1/jpg/fig-6-2.jpg*]

Investigation 9B:
METHANE HYDRATES: MAJOR IMPLICATIONS FOR CLIMATE

Driving Questions: *What are methane hydrates? What role might methane hydrates have played in causing the PETM (Paleocene-Eocene Thermal Maximum) and what were the possible impacts on atmospheric and ocean carbon dioxide levels? Are there similarities between PETM and current anthropogenic loading of carbon dioxide in the atmosphere and ocean?*

Educational Outcomes: To describe methane hydrates and how they can impact concentrations of carbon dioxide in the atmosphere and ocean. To examine any similarities between what happened during the PETM and the current upward trends in atmospheric and oceanic carbon dioxide.

Objectives: After completing this investigation, you should be able to:

- Describe the chemical and physical characteristics of methane hydrate and its distribution in the Earth environment.
- Demonstrate how methane hydrates could be major sources of atmospheric and oceanic carbon dioxide.
- Compare possible similarities in the role of methane hydrates in atmospheric and oceanic carbon dioxide concentrations during PETM and modern climate change.

PETM and Methane Hydrates

The PETM, the abrupt warming of Earth's atmosphere and ocean and associated environmental impacts that occurred about 55.8 million years ago, was global in scale and its impact lasted for more than 100,000 years. It was likely triggered by the rapid emission of carbon dioxide (CO_2), or by methane (CH_4), which chemically reacts with oxygen to produce CO_2 (and water). Possible sources of the huge amount of carbon that was necessary for producing the PETM are not known with certainty. It could have been from the release of methane (CH_4) from decomposition of hydrate deposits in seafloor sediments, CO_2 from volcanic activity, and/or oxidation of organic-rich sediments. There is considerable scientific evidence pointing to the possibility that the release of methane from naturally occurring solid methane hydrate deposits in ocean sediments played a prime role in producing the PETM.

Figure 1 shows a sample of methane hydrate. If ignited in air, it burns, as seen in the image. Up to 170 volumes of CH_4 as a gas at 1 atmosphere of pressure can be contained in one volume of methane hydrate. **Methane hydrate** (also called methane clathrate and methane ice) belongs to a

Figure 1. Burning methane hydrate. [National Research Council of Canada, NRC-SIMS]

unique class of chemical substances composed of molecules of one material forming an open solid crystal lattice that encloses, without chemical bonding, molecules of another material as represented in **Figure 2**. Methane hydrate is a solid form of H_2O that contains methane (CH_4) molecules within its crystal structure. It is the inclusion of sufficient methane molecules within the open cavities between water molecules that causes the stable solid structure to form (i.e., methane hydrate).

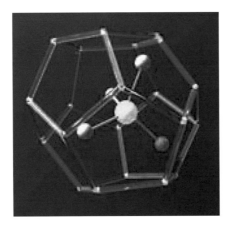

Figure 2.
Methane hydrate is a solid substance in which water molecules form an open crystalline lattice enclosing molecules of methane. [National Energy Technology Laboratory, DOE]

1. In Figure 2, the central molecule represents four (gray) hydrogen atoms bonded to a (yellow) [(*oxygen*)(*carbon*)] atom. The surrounding water molecules form the crystal lattice that makes the substance solid.

Within certain pressure and temperature ranges, methane hydrates form and remain stable in the Earth environment. To view methane hydrate formation and outcrops on the seafloor of the Gulf of Mexico and to answer the following questions, go to: *http://2100science.com/Videos/Frozen_Fuel.aspx*. Click on the arrowhead in the middle of the *YouTube* screen to start the program.

2. Early in the 8-minute video and again at about 7:40 minutes into the video you can view the burning of methane hydrate. A scientific error was made in the video narration with the statement that "methane hydrate is frozen methane." Evidence that this statement is probably not true is the [(*absence*)(*presence*)] of liquid water that can be seen as the burning of methane hydrate takes place. The flames seen result from the combustion of CH_4. The loss of CH_4 molecules leads to the collapse of the crystal lattices formed by the water molecules, so solid turns to liquid.

3. According to the narration (described at about 3:55), the actual seafloor observations in this video were made at a depth of [(*1200*)(*2100*)(*2900*)] feet.

4. The actual formation of methane hydrate can be seen in the video starting at about 4:50. The CH_4 entering the sampling tube in which the methane hydrate forms is in its [(*liquid*)(*solid*)(*gas*)] phase.

5. Methane hydrate outcrops can be viewed starting at 6:10. The existence of such outcrops is strong evidence that methane hydrate is [(*stable*)(*unstable*)] at the pressures and temperatures at that seafloor location.

Figure 3 is a methane hydrate phase diagram showing the combination of temperatures and pressures (given as water depth in meters). In the diagram "water-ice" refers to ordinary ice. Enclosed in the yellow area are combinations of temperature and pressure at which (solid) methane hydrate can stably exist.

6. The solid line between yellow and blue portions of the phase diagram marks the transition between conditions under which methane hydrate can or cannot exist. It indicates that at sufficiently high pressures, methane hydrate can exist at temperatures above the melting point of ordinary ice. At about 880 m (the depth of the Gulf of Mexico seafloor where the methane hydrate outcrops were observed in the above video), the temperature could have been as high as [(*+8 °C*)(*+10 °C*)(*+12 °C*)].

7. The phase diagram also shows that methane hydrate can exist at relatively shallow depths. At a temperature of 0 °C, methane hydrate could exist at water depths as shallow as about [(*300 m*)(*400 m*)(*500 m*)].

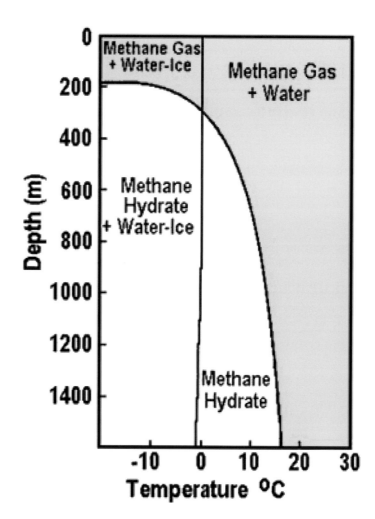

Figure 3.
Methane Hydrate Phase Diagram denoting depth (i.e., pressure) and temperature at which methane hydrate can exist. Methane hydrate is a solid. [National Energy Technology Laboratory, DOE]

It should be noted that the Figure 3 phase diagram is based on interactions of pure substances. In addition to temperature and pressure, the composition of both the water and the gas are critically important when making predictions of the stability of methane hydrates in different environments.

Methane Hydrates and the 2010 Gulf of Mexico Deepwater Oil Spill:

In the Deepwater Horizon oil spill catastrophe, methane hydrate was at least implicated as the possible source of the methane gas that likely caused the explosion and fire that destroyed the Deepwater Horizon oil drilling platform. In the initial clean-up efforts, it was unquestionably identified as responsible for thwarting the futile attempt of BP engineers to deploy an oil containment dome, acting as an upside-down funnel, to capture escaping oil and piping it to a storage vessel on the surface.

An oil containment dome was deployed on May 7-8 as one of the early attempts to cap the flow of escaping oil and natural gas (primarily composed of CH_4). It failed because the relatively low density methane hydrate that formed produced a slush that made the dome buoyant while at the same time clogging the pipe exiting at the dome's top.

8. The Figure 3 phase diagram shows why methane hydrate formed when methane and water mixed inside the containment dome. This can be seen by plotting a point on the phase diagram at a depth of 1500 m and temperature of 5.5 °C, representing the conditions at the seafloor well site. The plot falls within the [(*blue*)(*yellow*)] portion of the phase diagram indicating stable conditions for the existence of methane hydrate.

Methane Hydrates and PETM: Because the temperature and pressure ranges at which methane hydrates can exist are found throughout much of Earth's subsurface environment, it carries with it great potential to impact climate. This was as true in the past as it is now. As mentioned earlier, there is strong evidence that the release of CH_4 from methane hydrates might have been the primary forcing agent producing the PETM. This possibility has been thoroughly treated in learning materials developed by the Deep Earth Academy at: *http://www.oceanleadership.org/wp-content/uploads/2009/06/8_petm_abrupt-events.pdf*. We recommend that you examine the materials.

Global Distribution of Methane Hydrates: **Figure 4** describes the distribution of organic carbon in various Earth reservoirs. Gas hydrates are primarily methane hydrates, although other molecules of similar size to CH_4 (including hydrogen sulfide, carbon dioxide, ethane, and propane) form gas hydrates.

9. The one organic carbon reservoir greater than all the other reservoirs combined is [(*the ocean*)(*fossil fuels*)(*the land*)(*atmosphere*)(*gas hydrates*)].

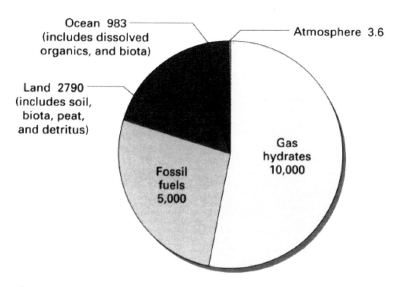

Distribution of organic carbon in Earth reservoirs (excluding dispersed carbon in rocks and sediments, which equals nearly 1,000 times this total amount). Numbers in gigatons (10^{15} tons) of carbon.

Figure 4.
Earth reservoirs of organic carbon.
[*http://marine.usgs.gov/fact-sheets/gas-hydrates/gas-hydrates-3.gif*]

Summary: **Figure 5** summarizes major aspects of methane hydrates in the Earth environment.

10. Figure 5 shows the two sources of the methane that are incorporated in the methane hydrate deposits. Biogenic generated CH_4 gas is the common by-product of bacterial ingestion of organic matter. It is considered to be the dominant source of the methane hydrate layers within shallow sea floor sediments. **[(*Hydrogenic*)(*Cryogenic*)(*Thermogenic*)]** generated CH_4 gas is produced by the combined action of heat, pressure and time on deep-buried organic material that also produces petroleum.

While not discussed here, large deposits of methane hydrates also occur in permafrost. With global warming and associated thawing of vast frozen land areas, the expectation is that significant quantities of CH_4 will be released. Their oxidation will lead to additional CO_2, enhancing the greenhouse effect.

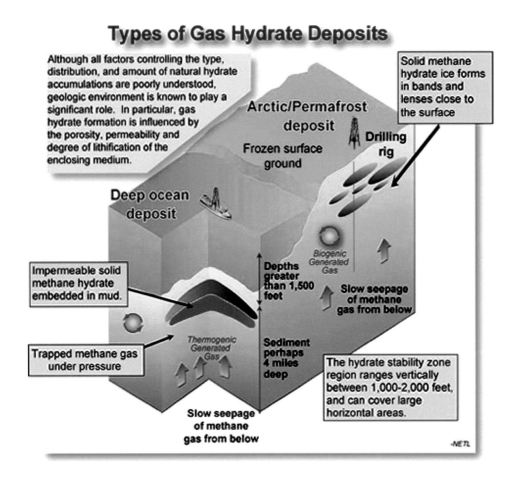

Figure 5.
Types of gas (mostly methane) hydrate deposits. [U.S. National Energy Technology Laboratory (NETL) *http://204.154.137.14/technologies/oil-gas/FutureSupply/MethaneHydrates/about-hydrates/geology.htm*]

CLIMATE AND CLIMATE VARIABILITY FROM THE INSTRUMENTAL RECORD

Driving Questions: *What can we determine from the instrumental record regarding the variability of climate? When might that variability indicate a change in climate?*

Educational Outcomes: Understanding climate and its variability are the first steps in making sense of what factors determine the mean state of the climate system, how it may have changed, and how it might change in the future. This course attempts to cover those key concepts of the boundary conditions that affect Earth's climate state and how the system varies within the limitations imposed by those constraints. We use the record of instrumental observations from the reliable length of readings to define that climate state and to identify its variability. From statistical analysis of the record, we then try to determine if and when a change has occurred or can be expected to occur in that climate state.

Objectives: After completing this investigation, you should be able to:

- Describe where climate data may be obtained and displayed.
- Show ways the climate record may be analyzed.
- Explain how climate analysis provides an understanding of climate variability and could lead to objective evidence of climate change (past and present).

The Instrumental Climate Record

The instrumental climate record is comprised of the recorded observations of temperature, precipitation, wind speed and direction, humidity, clouds, atmospheric pressure and the like taken from the beginnings of the invention of each instrument. However, a considerable period of time passed before each type of instrument was considered reliable. For example, Galileo Galilei (1569-1642) is credited with creating an instrument for measuring the changing temperature of the air. His mechanism consisted of a glass tube with a sealed bulb at one end that was warmed and the open end immersed in water. As the air volume in the tube cooled, water was drawn up into the tube. Subsequently, when the surrounding air temperature changed and the air in the tube came to equilibrium, the water level within the tube varied. This type of instrument really reacted to the sensible heat of the ambient environment. It is now termed a *thermoscope*.

Gabriel Daniel Fahrenheit (1686-1736) is credited with the invention of the true thermometer using a reliable liquid, mercury, in a narrow evacuated tube above a large bulb reservoir and based on a scale utilizing two fixed reference points. Only after the scientific community agreed on the reliability of the thermometer as a measuring device and reached consensus as to its representativeness in placement and use could the observations be considered comparable and valid. Because these actions came about through international conventions and the formation of national meteorological organizations, as well as the global spread of population, reliable widespread values of meteorological variables are considered to have been established by the late nineteenth century.

In this investigation, we will use temperature records to demonstrate ways in which climatic data acquired by use of reliable instrumentation are employed to examine climate and its variability. We will assess some specific temperature records for the period from 1895 to the present in the climatological division of Nebraska that includes Grand Island, whose data appear in Investigation 2A dealing with its local climatic data.

The temperature data we will analyze are monthly averages from the several sites within climatological division 5 ("central") of Nebraska. A *climatological division* is a region of a state considered to be homogeneous climatologically and containing a reasonable number of observing sites. **Here we will deal only with the July temperature averages;** each value thus being the average of 31 daily averages from approximately ten stations. The data are accessed from NOAA's National Climatic Data Center. These data represent the average temperature for what is normally the warmest month of the year in the area that includes and surrounds Grand Island, Nebraska for the reliable length of temperature readings.

The Nebraska climatological region 5 data were found from the course website, under the **Climate Information** section, subset Historical Information, "Climate Data Online" link. Then select *U.S. Divisional Data*. The resulting page provides a form to retrieve data by choosing, in this case, Division, State: Nebraska, Division: 05 – Central, Select Period Start: 07, 1895, Select Period End: 07, 2011, Static Graphs and Temperature, Show: July, then finally, **Submit**.

In statistics, there are several terms for the most representative or middle value of a series of data (including mean, median, and mode). *Average*, a sometimes vague term, is commonly used to denote the middle value of a set of data determined by dividing the sum of the values by the number of values used in the summation. This definition of average is synonymous with the term *mean,* which statisticians prefer using to precisely define a middle value so calculated. Finally, there is the term "normal" used in climatology. The climatological standard *normal* is the average (mean) of the 30 values of the variable measured within the most recent three-decade period, currently 1981-2010. Here we are using the average (mean) temperature for the month of July for each year. That average itself was the average of the daily average temperatures for the 31 days of each July. And each daily average was the average of that day's high and low temperatures.

Figure 1 is the time series graph of July average temperature in degrees Fahrenheit from 1895 to 2011 for the central Nebraska climatological division (supplied by NOAA's National Climatic Data Center). The average July temperature of each year is shown by a small green square plotted at the mid-point of that year with the years connected by line segments. The series of red squares making a horizontal line is the average (mean) of these July averages.

1. The average of all the July temperatures, denoted by the line of plotted red squares, is about **[(_74.5_)(_75.1_)(_77.2_)]** °F.

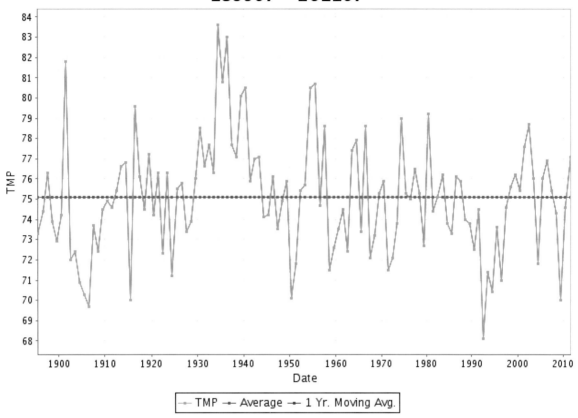

Figure 1.
Nebraska Central climatological division average July (month 07) temperatures from 1895 to 2011. TMP is abbreviation for temperature. [NOAA/NCDC]

2. Of the 117 years of average July temperatures reported in Figure 1, [(*1*)(*5*)(*10*)(*20*)] of the individual green squares appear(s) superimposed on a red square (i.e. only a green square appears that year). That is, the average of such an individual year is the same as the overall average.

3. In Investigation 2A the *Local Climatic Data, Annual Summary* publication for Grand Island, Nebraska, gave the average July temperature ("average dry bulb") for Grand Island in 2011 as 79.4 °F. From Figure 1, the 2011 central climatological division which contains Grand Island had an average July temperature that was [(*less than*)(*equal to*) (*greater than*)] the Grand Island value. The more urban nature of Grand Island compared to the rural area probably accounts for the difference.

4. The departure of the average July temperature of individual years from the long-term average is the variability. The maximum July average temperature in this series is about [(*75.1*)(*80.5*)(*83.6*)(*84.3*)] °F. This occurred in 1934.

5. The minimum July average temperature was 68.1 °F which occurred in [(*1906*)(*1933*) (*1971*)(*1992*)].

6. The *range* is defined as the difference between the maximum and minimum values, and is one simple measure of the variability of data. For the average July temperatures of this period, the range was [(*0.7*)(*8.4*)(*15.5*)(*29.3*)] F°.

7. Another issue in a long time series of data such as this one is the nature of variations in the data. For example, are there clear cycles of variation? If the values varied in a totally random way, one would expect swings from above to below average and vice versa occurring every year or at least every couple of years. One would not expect consistent runs of above or below average values and certainly not much above or below for extended periods. Using this general context, consider the periods 1902 to 1911, 1929 to 1943, and 1988 to 1997. Do these consecutive-year periods seem to imply total randomness in year-to-year variability? [(*Yes*)(*No*)]

Operating within Earth's climate system are many interacting processes, some of which serve to maintain certain departures (positive feedbacks) and some to dampen those departures (negative feedbacks.) Evidence exists in the Central Nebraska July temperature for both types of behavior.

The traditionally accepted climate period of the most recent three decades (30 years) of record for climatic values would dictate 1981-2010 as the current climatic normal period. Draw vertical lines on your graph at 1981 and at 2010 to display this span of years. Using the July average temperatures from these years, the following statistics were derived:

        mean: 74.2 °F
        maximum: 78.7 °F
        minimum: 68.1 °F

8. The 1981-2010 mean July temperature is [(*less than*)(*equal to*)(*greater than*)] the average of all July temperatures for 1895-2011.

9. The range of July temperatures for 1981-2010 is [(*4.9*)(*10.6*)(*14.5*)] F°.

10. The mean and range for the 1981-2010 period [(*are*)(*are not*)] equal to those for the 1895-2011 period.

In fact, one could compute the mean for each of the climatic normal periods within the entire length of this temperature record. Those mean values are given below:

1901-1930	74.6 °F
1911-1940	76.6 °F
1921-1950	76.4 °F
1931-1960	76.6 °F
1941-1970	75.2 °F
1951-1980	75.2 °F
1961-1990	74.9 °F
1971-2000	74.2 °F

A major concern facing humankind at present is the potential for global warming. This concern comes from the recent worldwide examination of trends in surface temperatures. We can consider these issues from our brief look at the localized summer temperatures of central Nebraska as a data "game". (By no means should this be considered a rigorous climate investigation but only a very limited example of how climate data are used to answer possible questions.)

11. Let us take "climate change" for our purpose here to imply simply a change of the mean temperature from one climatic normal period to another. Compare the mean of the most recent period, 1981-2010, to those of the preceding periods listed above. The 1981-2010 mean July temperature is [(*less than or equal to any*)(*greater than any*)] of the prior climatic periods listed.

12. If merely any change of mean value is considered a different climate regime, has there been a "climate change" over any of these periods? [(*Yes*)(*No*)]

13. Next, consider the magnitude of the difference from the others in the group and also recall the variability displayed in the graph of the total group of temperature values. Given the small difference of the 1981-2010 mean from those others and the overall nature and size of departures over the entire period of record (variance), could one conclusively be sure that real change had occurred? [(*Yes*)(*No*)]

14. If we are concerned about "global warming," compare the mean for the 1981-2010 period to those of the other periods listed. The 1981-2010 period mean July temperature is [(*tied with one other period with the lowest value*)(*lower than several others*) (*in the middle of the values*)(*higher than most others*)].

    One caution on these results, while there may be global trends, climate patterns are non-uniform geographically on the regional and local levels. Thus, while some areas may be clearly warming, others may be cooling, with only the overall average showing warmer values.

Another more definitive statistical procedure is to search for trends in data values. Is the time series of values of a climatological quantity uniformly increasing or decreasing and if so by how much? Clearly this may be difficult to detect from the graph alone given the variability displayed from year to year. The idea is to find a "line of best fit" to the entire series of data. While graphically placing a line is possible, different individuals may disagree on the placement. A rigorous statistical procedure exists however, called linear regression or least squares fitting of a line. Basically, the procedure minimizes the total sum of squared differences of each point from the line position. For further details, refer to standard statistical discussions.

The *trend* is given by the slope of the best fit line. If the slope is positive, values are generally increasing with time. A negative slope means values are generally decreasing. The magnitude of the slope value gives the rate of change, e.g. here in degrees per year. The slope values for the climatic periods are:

1901-1930	+0.08 F°/yr
1911-1940	+0.19 F°/yr
1921-1950	0.00 F°/yr
1931-1960	–0.19 F°/yr
1941-1970	–0.02 F°/yr
1951-1980	–0.01 F°/yr
1961-1990	–0.01 F°/yr
1971-2000	–0.06 F°/yr
1981-2010	+0.02 F°/yr

15. Consider the values of the trends for each climatic period. These July average temperature trends are [(***all decreasing***)(***mixed with both increases and decreases***)(***all increasing***)].

16. The greatest trend value (greatest change in magnitude, whether positive or negative) was during the periods [(***1901-1930 and 1941-1970***)(***1911-1940 and 1931-1960***)(***1921-1950 and 1981-2010***)].

The maximum average July temperature in central Nebraska occurred in 1934. This was the infamous "Dust Bowl" period of severe drought in the central U.S. brought on by persistent dry conditions which aided in solar heating becoming predominantly sensible heating, resulting in a large Bowen ratio. The 1911-1940 period ending with the Dust Bowl time is reflected in that period's trend. Perhaps as remarkable is the obvious trend from about 1934 to 1950 of declining temperatures. This does not show in the values above due to the placement across the arbitrarily chosen climatic periods.

Let us now consider the recent attention to the global warming issue. From Figure 1, a warming trend appears to fill much of the most recent time, since 1992. Is this evidence of the much discussed anthropogenic influence on climate? The following statistics can be found for values from two selected, nearly decadal periods:

	1993 – 2002	2001 – 2009
mean:	74.5 °F	75.2 °F
maximum:	78.7 °F	78.7 °F
minimum:	70.4 °F	70.0 °F
trend:	+0.86 F°/yr	–0.66 F°/yr

17. In terms of the mean and extremes, does there appear to be any substantial difference between these two chosen periods? [(*Yes*)(*No*)]

18. The variation in the trend produced by the selective inclusion of the years of data does greatly affect perceived trend statistics however. For which period would the trend suggest a more alarming warming swing? [(*1993-2002*)(*2001-2009*)].

19. Compare these two period trends, 1993-2002 and 2001-2009, to those 30-year climatological period trends listed above. The 1993-2002 trend is [(*much less than*)(*approximately equal to*)(*much greater than*)] the largest trend value of the prior climatic periods listed.

Therefore, overall recent conditions (1993-2011, the present), at least in Nebraska's central climatological division, have not been the largest change of temperature that were experienced. And the temperatures experienced were not the warmest seen in the record. In fact, this period from 1993 to 2011 contained one of the lowest July average temperatures recorded in central Nebraska. It should be clear that selection of historical values must be carried out using standard statistical procedures as a guide and not arbitrarily to produce a desired effect.

One last issue; the overall trend for the entire 1895-2011 period gives a value of 0.00! [In Figure 1, such a trend line would be superimposed on the red average line.] In other words, from this particular instrumental temperature record, there is no demonstrable trend of increasing summer temperature values. But, too, this is only one set of data from one climatological division in Nebraska.

Summary: A variety of ways in which recorded temperature values can be analyzed has been demonstrated. Although limited to July temperatures in one climatological division, they are intended as examples of how empirically acquired climatic data can be examined to determine climate variability, which in turn can lead to objective evidence of climate stability or climate change in the past and present.

While statistical analyses inform past and current climate, predicting climate change in the future requires a dynamic approach. As mentioned in Investigation 1A, while the empirical approach allows us to construct descriptions of climate, the dynamic approach is what enables us to seek explanations for climate. Explanations lead to predictions, including prognostications of change. Considering Earth's climate system dynamically as a physical system makes it possible to leap-frog into the future via computer models to make potentially useful climate change predictions.

The USGCRP *Global Climate Change Impacts in the United States* report presents climate-change predictions on the regional scale. As an example, the report shows in **Figure 2** that climate models predict that Nebraska's central climatological division's summer temperatures will rise 6 F° or more compared to a 1960s and 1970s baseline. And recall from Investigation 2A that the 1981-2010 state-wide annual temperature maximums and minimums have increased.

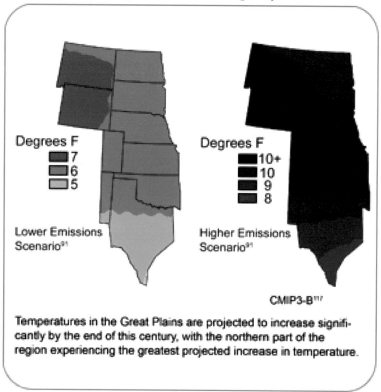

Figure 2.
Projected summer temperature change in the Great Plains. [USGCRP].

Investigation 10B: RICE GROWING AND CLIMATE CHANGE

Driving Question: *How might large-scale climate change affect agriculture in a specific region?*

Educational Outcomes: To describe the possible impact of higher temperatures associated with climate change on the heat intolerance of rice growing. Demonstrate that both temperature means and variabilities (spreads of temperature values about their means) can impact rice production. To explain that other climate-change factors can adversely impact rice and other food production.

Objectives: After completing this investigation, you should be able to:

- Demonstrate how climate might change in the future.
- Describe the possible implications of increased variability in climate.
- Explain how agricultural productivity might be vulnerable to climate change.

Climate Change Impacts of Agriculture: A Case Study

Global-scale climate change is likely to affect all sectors of society, including agriculture and food production. To understand the potential impact of climate change on agriculture, we need to consider the ways in which climate might change in the future. The climate record indicates that climate change may consist of long-term trends in average values of weather measures (i.e., temperature, precipitation), changes in the frequency of weather extremes, or both.

This investigation focuses on how higher July temperatures, more variable July temperatures, or both, might impact the cultivation of rice in southern China. Rice production is highly climate sensitive, most notably in terms of temperature (and also because about 75% of the world's production involves irrigation). We will use a simple climate model for average daily July temperatures at Nanning (22.6 degrees N, 108.2 degrees E), located in a rice-growing area of southern China with a tropical monsoonal climate.

The production of rice is extremely important. It is the staple food for almost half of the world's population, especially in tropical Latin America, East, South and Southeast Asia (**Figure 1**). It is the grain with the second highest worldwide production, after maize ("corn") according to the U.N. Food and Agricultural Organization. Since large proportions of maize crops are grown for other than human consumption, rice is probably the most important grain with regards to human nutrition and caloric intake.

It is estimated that in 2004, 600 million metric tons of paddy rice were produced world wide. (Final milled rice is about 68% of the weight of paddy rice.) That year the top producer was China, which had about 26% of the world's production. India with 20%, Indonesia at 9% and Bangladesh followed. In 2011, 653 million tons of paddy rice was produced according to the United Nation's Food and Agriculture Organization (FAO).

Figure 1.
Worker transplanting rice seedlings in Korea. (IRRI Images)

A. Temperature Requirements for Rice

Recent research suggests the possibility that the heat intolerance of rice may make it difficult to maintain or increase production of this food staple. Studies of the growth of two common species of rice at elevated temperatures, consistent with proposed projections of temperature change, indicated that the fertility of rice spikelets (flower shoots that become the seeds, i.e. edible rice grains) may be at risk. Spikelet flowering at a control temperature of 29.6 °C, was compared to that at 33.7 °C and 36.2 °C. Results showed a 7% reduced spikelet fertility per degree rise above 29.6 °C in one species, and a 2.4% reduction per degree in the other species with increasing rate of reduction as temperatures rose. Both species became sterile when the temperature reached 33.7 °C at flowering (*http://jxb.oxfordjournals.org/cgi/content/abstract/erm003v1*). A recent study at the International Rice Research Institute (IRRI) has shown that each one degree Celsius increase in the minimum temperatures at night can decrease rice yield by about 10 percent.

Adaptations of rice species or its environment will be needed if the higher extremes of Intergovernmental Panel on Climate Control (IPCC) projections become reality. The IPCC models imply that global average temperatures may be from 1 to 6 C° higher by the end of the 21st century.

For the Northern Hemisphere, July on average is the warmest month of the year, making it the month when crops are at the greatest risk to stress caused by episodes of excessively high temperatures. As already indicated, the temperatures at flowering time for planted rice are crucial in the development of the grains and thus the total crop produced.

Figure 2.
Nanning is located in major rice growing area of southern China. [Adapted from CIA World Factbook]

We will consider Nanning as it is in the world's greatest rice growing country and the most populous country on the planet (see **Figure 2** map, and **Figure 3** view of Nanning-area rice fields). The control temperature used in the rice study quoted is also near the average monthly high temperature for Nanning based on the temperature record for the 1951-2008 period. The average daily temperature, the average of the high and low for the day, provides the data series employed in our climate model. The series of average July monthly temperatures from 1951 to 2008 in Nanning yields an average of 28.4 °C. This value is near the experimental study's rice control value (29.6 °C) and less than the values shown to be harmful under warmer conditions. For simplicity, we will consider our conclusions for the model based on the Nanning temperature series average.

Figure 3.
Workers in the rice fields, Nanning, China. [*http://www.travelpod.com/travel-photo/ksshepard/1/1238117400/workers-in-the-rice-fields.jpg/tpod.html*]

B. 1. Nanning Climate Model

This investigation utilizes a simple statistical climate model of Nanning's July average daily temperatures, displayed graphically in **Figure 4**. The horizontal axis is the temperature in degrees Celsius and the vertical axis is the probability of a particular temperature's occurrence. The probability of occurrence increases upward. **Curve A** represents the frequency of occurrence of daily average July temperatures during the present climatic regime. You may recognize this model distribution as "normal" (also termed bell-shaped or Gaussian). This model was used in Investigation 2B.

Based on the 58-year record, the Nanning mean daily temperature in July is 28.4 °C. Further analysis of the data employed in determining Curve A indicates that about two-thirds of all average daily temperatures occur within 0.6 Celsius degree of 28.4 °C (plus or minus one standard deviation), that is, between 27.8 °C and 29.0 °C. Assuming no change in climate, Curve A can be used to estimate the relative frequency of any average daily temperature occurring during July at Nanning.

1. Curve A depicts the probability of occurrence curve of average July daily temperatures. It shows that the most frequently (most likely) occurring average daily temperature in July is [(*26.2*)(*28.4*)(*28.8*)] °C. Draw a vertical straight line on the graph at this temperature value.

2. Assuming the rice tolerance study described earlier is applicable, temperatures that were above 29.6 °C resulted in reduced rice growth. On the same graph, draw a vertical line at 29.6 °C to depict the harmful threshold. According to Curve A, and the position of the line you just drew, there [(*is*)(*is not*)] a possibility that under current climatic conditions a July could occur with a high enough daily average temperature to reduce rice growth.

3. The same rice tolerance study showed that both rice species studied became sterile when the temperature reached 33.7 °C during flowering. Draw a vertical straight line at that value on the graph. According to the graph, it [(*is likely*)(*is highly unlikely*)] that under current climatic conditions a July could occur with a high enough daily average temperature to make the flowers sterile.

B. 2. Nanning Climate Model – 4 C° Warmer Than Present

Recent runs of global climate models predict that the global mean annual temperature could rise between 1 and 6 Celsius degrees by the end of this century. **Let us assume in this investigation that this global-scale warming translates into a warming of 4 Celsius degrees in Nanning's July mean daily temperature.** Let us also assume initially that the variability of July average daily temperatures remains the same; that is about two-thirds of all values will occur within ±0.6 Celsius degree of the new July mean daily temperature.

4. The new July overall mean daily temperature is 32.4 °C (28.4 °C + 4 Celsius degrees = 32.4 °C). In the warmer climate that experienced no change in variability, about two-thirds of all mean daily temperatures would be expected to fall between 31.8 °C and [(*33.0*)(*33.4*)(*33.8*)] °C.

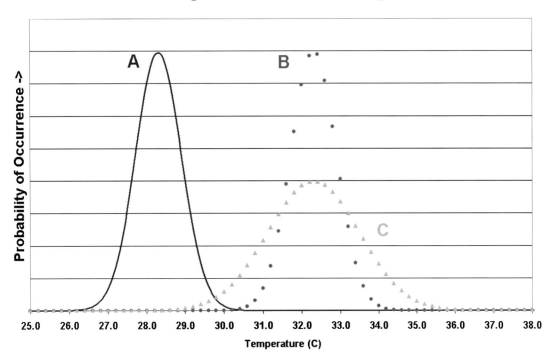

Figure 4.
Climate model of Nanning, China July mean daily temperatures.

5. In Figure 4, connect the set of small red dots (plotted at a 0.2 C° interval) with a smooth line to produce a new model distribution (identical in shape to Curve A) representing the warmer July climate. The curve is already labeled with a red "B". Draw a vertical straight line at 32.4 °C representing the new July mean daily temperature. The peak of **Curve B** (the highest probability of occurrence) is [(*the same as*)(*different from*)] the highest point on Curve A indicating that the likelihood of the respective mean daily temperature is the same for both models.

6. Both curves have the same width indicating that the variability or spread of temperatures about the mean is [(*the same*)(*different*)] for both models.

7. The area under each curve is directly proportional to the total number of occurrences of the temperatures represented. Compare Curve B with the position of the 29.6 °C harmful threshold line you drew. The comparison shows that the warming would reduce rice production in the Nanning area during [(*some*)(*most*)(*essentially all*)] years.

8. Because of the 4 C° warming, the rice species in the research study that is more temperature sensitive (7% reduced spikelet fertility per degree above threshold) would likely see an average of about [(*7%*)(*20%*)(*28%*)] reduction in spikelet fertility.

9. According to Curve B and the position of the line you drew showing the temperature (33.7 °C) when both rice species during flowering become sterile, there [(*is*)(*is not*)] a possibility that under the higher temperatures a July could occur with a high enough daily mean temperature to totally decimate rice production via those species.

B. 3. Nanning Climate Model – Warmer and More Variable

10. Now assume that in addition to the 4 Celsius-degree rise in the July daily mean temperature, the new climatic regime features much greater variability (spread) in daily mean temperatures. In this new model, the variability about the mean is arbitrarily doubled so that about two-thirds of all values will occur within ±1.2 Celsius degrees of the overall mean July daily temperature of 32.4 °C, that is, between [(*30.0*)(*31.2*)(*31.8*)] °C and 33.6 °C.

11. In Figure 4, connect the set of small green triangles (also plotted at a 0.2 C°-interval) with a smooth line to graphically represent the warmer and more variable July climate. This has already been labeled curve "C". **Curve C** is broader than Curves A and B indicating [(*increased*)(*decreased*)] variability of average daily temperature.

12. The high point of Curve C is [(*lower than*)(*equal to*)(*higher than*)] the high points of Curves A and B.

13. This indicates that there is a(n) [(*reduced*)(*increased*)] probability of occurrence of a daily average temperature equal to the peak mean daily temperature for July.

14. The area under Curve C to the right of the 33.7 °C (temperature at which both rice species during flowering become sterile) vertical line you drew on the map is [(*greater than*)(*less than*)(*the same as*)] the area under Curve B to the right of 33.7 °C.

15. This comparison indicates that with a 4 Celsius-degree increase in July mean daily temperature ***and*** an increase in temperature variability, the probability of occurrence of lethal conditions for rice production of these species would [(*increase*)(*decrease*) (*not change*)].

Note that the comparison of Curves B and C indicates the increased variability depicted by Curve C can also result in higher production of rice due to the increase in the probability of more years experiencing cooler July temperatures. There could even be a few years when the July daily mean temperature is below the 29.6 °C harmful threshold!

16. This investigation indicates that an evaluation of the potential impact of climate change on agriculture must take into account possible changes in climatic [(*averages*) (*variability*)(*both of these*)].

Summary: Climate change can result in major, perhaps devastating, impacts on agricultural productivity. Scenarios concerning rice production show that literally hundreds of millions of people are at risk in the coming decades due to reduced rice production. Temperature change is only one factor affecting rice production. Rising sea level, reduced water resources, and potential pests in rice-growing areas are also of grave concern.

For a detailed report published on 17 October 2011, "Rice Production and Global Climate Change: Scope for Adaptation and Mitigation Activities", by the International Rice Research Institute, go to: *http://irri.org/climatedocs/presentation_Lists/Docs/12_Wassmann.pdf*.

11A - 1

Investigation 11A: VOLCANISM AND CLIMATE VARIABILITY

Driving Questions: *How does volcanism influence climatic conditions? How long is its presence felt?*

Educational Outcomes: To describe the impacts of volcanic eruptions on the amounts of solar radiation reaching Earth's surface, especially in terms of resulting temperature variability. To determine how volcanoes cause temperature departures from long-term mean values and for how long. To explain how proxy data have been employed to relate historical explosive volcanic eruptions to temperature variability.

Objectives: After completing this investigation, you should be able to:

- List several natural forcing agents and mechanisms of Earth's climate system.
- Describe how one mechanism, volcanic activity, affects the system.
- Show how a series of proxy data can be used to explain observed climate variability.

Impacts of Volcanism on Climate Variability

Climate variability is the fluctuation of climatic elements such as temperature about the long-term mean value. The important question is: When does a difference from the climatic mean value, usually based on the average of the past three complete decades, constitute a sustainable change from the previous state and when is it just the natural variability of a statistical value about its constant mean? The boundary conditions of Earth's climate system are themselves variable forcings that drive the system to achieve slightly different states from one year to the next. We must be guided by the science and mathematics of such systems to discern through statistical analysis the level of confidence in deciding that real climate change has occurred.

The most significant boundary condition for Earth's climate system is the energy input to the system, energy received from our Sun. Although there are slight variations in the intensity of solar energy received at Earth's orbit from one year to the next and noticeably over the 11-year sunspot cycle, for most purposes we can consider this input to be relatively constant.

One sometimes significant variation of this energy input to the Earth system occurs when volcanoes eject large quantities of ash and gases into the atmosphere. Ash particles, if fine enough in size and shot high enough into the atmosphere, can have fairly long lifetimes aloft. The gases erupted, particularly sulfur dioxide, can combine with atmospheric water vapor to form sulfuric acid and then become solid sulfate aerosols in the stratosphere. The stratosphere (the atmospheric layer above the troposphere) is an extremely stable layer for air motions and very dry without precipitation. Particles at stratospheric altitudes, typically 11 to 18 km (7 to 11 mi) or higher, have residence times measured in years. **Figure 1** shows the dispersal of fine ash and sulfur dioxide over about a two-month period from the June 1991 massive volcanic eruption of Mount Pinatubo in the Philippines as demonstrated by a NASA Mount Pinatubo Particle Model. The eruption spewed materials that spread into the global

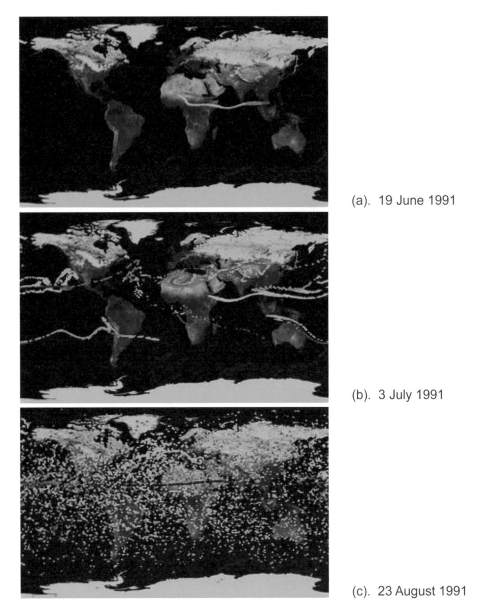

(a). 19 June 1991

(b). 3 July 1991

(c). 23 August 1991

Figure 1.
Global spreading of sulfur dioxide and dust from June 1991 Mount Pinatubo eruption over about a two-month period following the eruption. [NASA Scientific Visualization Studio]

stratosphere and contributed to a –0.3 C° Northern Hemisphere temperature anomaly in the following year, 1992.

Observe an animation of the NASA Mount Pinatubo Particle Model by going to: *http://svs.gsfc.nasa.gov/vis/a000000/a002300/a002389/*. To animate, select one of the animation formats listed to the right of the top image (e.g., MPEG-1). In both Figure 1 and the animation, red is high and blue is closer to Earth's surface.

1. The Mount Pinatubo Particle Model shows that the initial atmospheric flow of erupted fine ash and sulfur oxide was from [(***west to east***)(***east to west***)]. Subsequently, the prevailing westerlies dispersed the material to higher latitudes.

2. The Mount Pinatubo Particle Model shows that within two months of the 1991 event, eruption products from the point source at 15.1 degrees North latitude spread around the globe [(*only in the tropical latitudes*)(*only in the Northern Hemisphere*) (*from tropical to polar latitudes in the Northern and Southern Hemispheres*)].

The fine ash and sulfur dioxide from violent volcanic eruptions such as the 1991 Mount Pinatubo episode can block incoming solar radiation, thus decreasing the energy available at Earth's surface for atmospheric warming. In this way the year or so following extremely violent volcanic explosions can be cooler than normal across much of the planet.

Eventually air motions will mix these particles into the lower troposphere where atmospheric circulation, precipitation, and the constant pull of gravity will bring them back to Earth's surface. Although recorded history of such volcanic activity is relatively short, layers of volcanic dust and sulfur particles found in sediment and glacial ice cores can be sources of information on ancient volcanic eruptions and proxy climate data. **Remarkably, the volcanic eruption record even appears in tree growth rings!**

Tree Rings as Temperature Proxies: Trees in middle and high latitudes react to the dominant annual cycle in solar radiation by growing and adding a layer of new wood each year. It has long been known that for trees located in marginal habitats (locations that support only a few species or individuals because of limiting environmental conditions), the width of the annual ring of growth is related to the temperature and precipitation of that season. Thus a cross-section from felled trees or even a small diameter core bored from a living tree can provide clues to the sequence of climate conditions to which the tree was subjected during its growth. Considerable success has been achieved in using tree growth ring records for reconstructing the chronology of past droughts. Of course there are many subtleties involved in tree ring interpretation. For more information, see: *http://www.ltrr.arizona.edu/dendrochronology.html*.

A more recent insight has been within the tree rings' structure itself. For each yearly ring, there are two types of wood, a lighter, less dense early wood layer and a darker, denser latewood section. According to the latest findings, the density of the latewood portion of each layer is most closely correlated to the temperature of that year. To account for variations in location, age of the tree, etc., the studies are typically done by creating anomalies for each year of the sequence. This process has been carried out for many trees in the marginal Northern Hemisphere boreal forest where the growing season is short and particularly sensitive to climatic variability.

One researcher in this area, Keith Briffa with the Climatic Research Unit of the University of East Anglia (UK), has done extensive work relating northern tree rings to temperature. His dataset is available from NCDC's World Data Center (WDC) for Paleoclimatology linked from the course webpage under **Climate Variability**, Climate Forcing Data. Click on this link and scroll down to the list under Volcanic Aerosols, then click on "*Major Volcanic Eruptions, 600 Years, Briffa et al. 1998*". There, you will find Northern Hemisphere temperature reconstructions of dimensionless yearly sequences of tree ring latewood density anomalies and one sequence of Northern Hemisphere temperature anomalies. The time series values are for the years from 1400 to 1994. Such series of values would be useful to anyone who wishes to study variations of these quantities over time.

Below the time sequences are two summary tables of data from the WDC website. **Table 1** is a ranking of the most negative 5% of the tree ring density anomalies and the corresponding temperature anomalies including year of occurrence.

Table 1. Ranking of the Northern Hemisphere most negative 5% of tree-ring anomalies (NHD1) and corresponding temperature anomalies (NH anom. C degrees).

Rank	Year	NHD1 sValue	NH anom. degrees (C)
1	1601	−6.90	−0.81
2	1816	−4.33	−0.51
3	1641	−4.31	0.5
4	1453	−4.24	−0.5
5	1817	−3.76	−0.44
6	1695	−3.50	−0.41
7	1912	−3.33	−0.39
8	1675	−3.13	−0.37
9	1698	−3.08	−0.36
10	1643	−2.99	−0.35
11	1699	−2.96	−0.35
12	1666	−2.89	−0.34
13	1884	−2.89	−0.34
14	1978	−2.80	−0.33
15	1837	−2.78	−0.32
16	1669	−2.77	−0.32
17	1587	−2.64	−0.31
18	1740	−2.61	−0.3
19	1448	−2.57	−0.3
20	1992	−2.56	−0.3
21	1836	−2.48	−0.29
22	1818	−2.45	−0.29
23	1495	−2.42	−0.28
24	1968	−2.38	−0.28
25	1742	−2.35	−0.27
26	1783	−2.35	−0.27
27	1667	−2.35	−0.27
28	1642	−2.22	−0.26
29	1819	−2.21	−0.26
30	1446	−2.20	−0.26

3. The latest volcanic eruption evidenced in the list (which occurred at Pinatubo, Philippines) was in 1991. According to the table, its Northern Hemisphere tree-ring anomaly value (listed for 1992) of –2.56 corresponded to a Northern Hemisphere temperature anomaly of [(*–0.26*)(*–0.28*)(*–0.3*)] °C.

4. According to the Table 1 listing, the [(*1600s*)(*1700s*)(*1800s*)(*1900s*)] were subjected to the greatest negative tree-ring anomalies and corresponding temperature anomalies.

Table 2 is a listing of the largest explosive volcanic eruptions since 1400 including the year, name of volcano, location by latitude and longitude.

Table 2. Largest explosive volcanic eruptions since AD 1400.

Year	Season	Volcano and region	Lat. (deg)	Long. (deg)	VEI
1450	–	Aniakchak, Alaska	56.9N	158.1W	5(?)
1452	–	Kuwae, Vanuatu, SW Pacific	16.8S	168.5E	6
1471	3yr	Sakura-Jima, Japan	31.6N	130.7E	5(?)
1477	1	Bardarbunga (Veidivotn), Iceland	64.6N	17.5W	5(?)
1480	–	St Helens, Washington, US	46.2N	122.2W	5(+)
1482	–	St Helens, Washington, US	46.2N	122.2W	5
1580	–	Billy Mitchell, Bougainville, SW P	6.1S	155.2E	6
1586	–	Kelut, Java	7.9S	112.3E	5(?)
1593	–	Raung, Java	8.1S	114.0E	5(?)
1600	1	Huaynaputina, Peru	16.6S	70.9W	6(?)
1640	3	Komaga-Take, Japan	42.1N	140.7E	5
1641	1	Parker, Philippines§	6.1N	124.9E	6
1660	–	Long Island, New Guinea	5.4S	147.1E	6
1663	3	Usu, Japan	42.5N	140.8E	5
1667	4	Shikotsu (Tarumai), Japan	42.7N	141.3E	5
1673	2	Gamkonora, Halmahera	1.4N	127.5E	5(?)
1680	–	Tongkoko, Sulawesi	1.5N	125.2E	5(?)
1707	1	Fuji, Japan	35.4N	138.7E	5
1739	3	Shikotsu (Tarumai), Japan	42.7N	141.3E	5
1800	1	St Helens, Washington, US	46.2N	122.2W	5
1815	2	Tambora, Lesser Sunda Is	8.3S	118.0E	7
1835	1	Cosiguina, Nicaragua	13.0N	87.6W	5
1853	1	Chikurachki, Kurile Is	50.2N	155.0E	5(?)
1854	1	Sheveluch, Kamchatka	56.7N	161.4E	5
1883	3	Krakatau, west of Java	6.1S	105.4E	6
1886	3	Okataina (Tarawera), New Zealand	38.1S	176.5E	5
1902	4	Santa Maria, Guatemala	14.8N	91.6W	6(?)
1907	2	Ksudach, Kamchatka	51.8N	157.5E	5
1912	1	Novarupta (Katmai), Alaska	58.3N	155.2W	6
1932	2	Azul, Cerro (Quizapu), Chile	35.7S	70.8W	5(+)
1956	2	Bezymianny, Kamchatchka	56.0N	160.6E	5
1980	2	St Helens, US	46.2N	122.2W	5
1982	2	El Chichon, Mexico	17.4N	93.2W	5
1991	3	Pinatubo, Philippines	15.1N	120.4E	6

In Table 2, VEI (Volcanic Explosivity Index) is a measure of the explosive strength of a volcanic eruption based on amount of ejecta and plume height with increasing numbers corresponding to increasing strength. The season of the eruption (where known) is denoted by a digit, 1 = winter (Dec-Feb), 2 = spring (Mar-May), 3 = summer (Jun-Aug), and 4 = fall (Sep-Nov).

5. According to the VEI rating listed in Table 2, the most explosive eruption of the period given was Tambora in the year [(*1482*)(*1600*)(*1815*)]. The worldwide effect of this event is chronicled in the *AMS Climate Studies* text as "the year without a summer".

6. From Table 2, the most frequently erupting volcano was [(*St. Helens*)(*Pinatubo*)(*Krakatau*)].

Table 3 employs data from Tables 1 and 2 to search for patterns between major explosive volcanic eruptions and negative (cold) Northern Hemisphere temperature anomalies as related to tree ring density anomalies.

Table 3 is constructed so the Eruption Years listed in the left column are the same as those listed in Table 2. These Eruption Years are the dates of occurrence of the 34 largest explosive volcanic eruptions in the time period 1450-1991. Dates (in italics) plotted in Table 3 were acquired from Table 1 which provided data on the years with the 30 most negative tree-ring and corresponding temperature anomalies. These italicized dates were plotted when the first anomaly year matched the year of the eruption or the year following the eruption. Subsequent anomaly years are plotted when there is no skip in years following the initial anomaly year.

7. According to Table 3, a total of [(*3*)(*4*)(*5*)] cold temperature anomalies occurred in the same year as a major eruption.

8. According to Table 3, a total of [(*3*)(*5*)(*9*)] cold temperature anomalies appeared for the first time in the year following the year of major eruption ("Year + 1").

9. Of the 34 greatest explosive volcanic eruptions listed in Table 3, about [(*15*)(*25*)(*35*)] % of them were followed by at least one of the most negative 5% of tree-ring anomalies and corresponding temperature anomalies as listed in Table 1 within the same year or one year following a major volcanic eruption.

10. Therefore, one can conclude that it is [(*likely*)(*not at all likely*)] that volcanic particles in the atmosphere from major explosive eruptions could result in blocking sunlight and lowering surface air temperatures.

11. Some of the eruptions had a fairly long-lasting effect. For instance, Tambora in 1815 was acknowledged as the most violent eruption in recent history with VEI = 7. Table 3 shows that Tambora's eruption probably led to significant cold temperature anomalies for [(*2*)(*3*)(*4*)(*5*)] subsequent years.

Table 3. Eruption year and subsequent year(s) with cold temperature anomaly.

Eruption Year	Year of Cold Temperature Anomaly				
	Same Yr	Yr + 1	Yr + 2	Yr + 3	Yr + 4
1450					
1452		*1453*			
1471					
1477					
1480					
1482					
1580					
1586		*1587*			
1593					
1600		*1601*			
1640		*1641*	*1642*	*1643*	
1641	*1641*	*1642*	*1643*		
1660					
1663					
1667	*1667*				
1673					
1680					
1707					
1739		*1740*			
1800					
1815		*1816*	*1817*	*1818*	*1819*
1835		*1836*	*1837*		
1853					
1854					
1883		*1884*			
1886					
1902					
1907					
1912	*1912*				
1932					
1956					
1980					
1982					
1991		*1992*			

12. Another way in which a lasting atmospheric cooling effect may occur is shown in the Table 1 rankings by the years 1641, 1642, and 1643. These years coincided or followed the eruptions of [(*Komaga-Take*)(*Shikotsu*)(*Sakur-Jima*)] in Japan in 1640 and Parker in the Philippines in 1641. It is likely that both eruptions simultaneously contributed to atmospheric cooling for at least two years.

In fact, cooling by volcanic sulfur dioxide and fine ash in the upper atmosphere may have resulted from other periods of significant and/or sustained but less dramatic eruptions. For example, 1968 is number 24 in rank, yet does not follow a listed volcanic explosion. However, Mt. Agung on Bali in Indonesia erupted violently in 1963-64, killing 1700 people. It was a massive sulfur-rich episode and was noted for its stratospheric aerosol input.

Summary: The primary external boundary condition of solar energy input to Earth's climate system can be impacted significantly by internal processes in the climate system that provide variability to the resulting annual state. One significant variation in the amount of this energy absorbed into the Earth system occurs when volcanoes eject large quantities of ash and gases into the atmosphere producing particles that reflect and scatter solar energy which would otherwise wend its way through the climate system. In total, Earth's climate system is the sum of all such processes and therefore exhibits inherent variations on scales of years to decades and millennia.

SNOW AND ICE ALBEDO FEEDBACK IN EARTH'S CLIMATE SYSTEM

Driving Questions: *What impacts do snow and ice albedo have on climate? Is the amount of snow and ice on Earth decreasing?*

Educational Outcomes: To describe the seasonal and permanent snow and ice cover impacts on global and regional climates. To explain how the presence or absence of snow and ice cover controls patterns of heating and cooling over Earth's surface more than any other surface feature. To examine observations that show a global-scale decline of snow and ice for many years, including the polar amplification of surface temperatures in the high latitudes of the Northern Hemisphere.

Objectives: After completing this investigation, you should be able to:

- Describe polar amplification and the observed changes in snow and ice cover in recent decades.
- Explain the concepts of positive and negative feedback and provide examples of such feedbacks resulting from different Earth surfaces including snow cover, ice, bare ground, and open water.

Polar Amplification: Observational data of Earth surface temperatures have shown an unequivocal pattern of planetary warming. However, this warming trend has not been uniform around the globe. **Figure 1** traces temperature change for three latitude bands (Northern, Low, and Southern Latitudes) from 1900 through 2011, based on their average annual global mean surface temperatures for the period of 1951-1980. Note the different temperature anomaly scales in the three graphs. In the individual Figure 1 graphs, black data points connected by dotted lines are actual temperature anomaly curves. The red curves are based on five-year running averages, and the green bars show uncertainty estimates.

1. All latitude bands in Figure 1 show long-term temperature anomaly trends, especially since the 1970's, indicating **[(*cooling*)(*steady temperatures*)(*warming*)]**.

2. Based on the red five-year running mean curves from 1900 to 2009, the greatest warming occurred in the **[(*northern-*)(*low-*)(*southern-*)]** latitude band.

The IPCC (*Climate Change 2007*) reported that during essentially the same time period, the average surface temperature in the Arctic increased at almost twice the global rate. This greater temperature increase in the Arctic is termed **Polar Amplification**. The term is generally not applied to the Antarctic as it has not experienced such a dramatic increase.

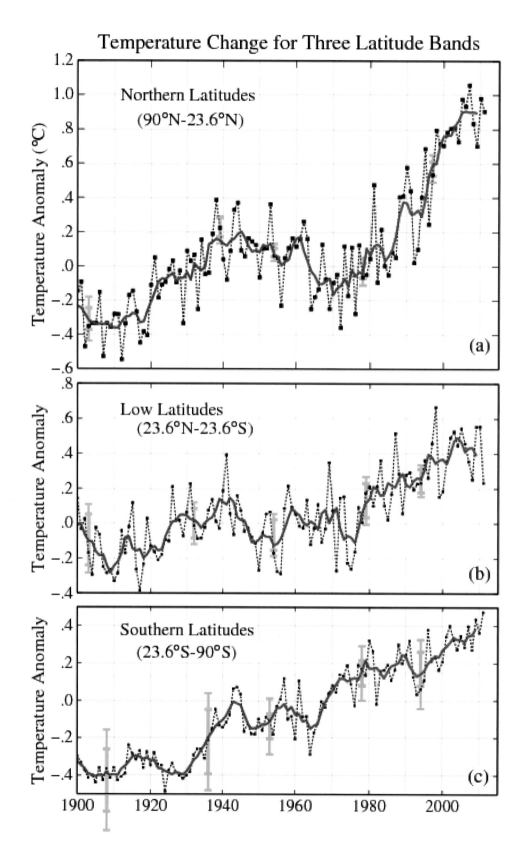

Figure 1.
Temperature departures (anomalies) from the long-term global mean surface temperature by latitude bands, 1900-2011. [Goddard Institute for Space Studies/NASA]

Figure 2 is an IPCC (*Climate Change 2007*) depiction of surface temperature anomalies at high latitudes and associated snow and ice conditions.

3. Compare Figure 2's surface air temperature anomaly graphs (A) north of 65 degrees N and (G) south of 65 degrees S. They show that in recent decades the increase in surface air temperatures was significantly greater in the [(*Arctic*)(*Antarctic*)].

4. Figure 2's graphs depicting Northern Hemisphere anomalies of (B) sea ice extent, (C) frozen ground extent, and (D) snow cover extent for recent decades indicate trends of [(*increasing*)(*steady*)(*decreasing*)] values. These are expected outcomes of Polar Amplification.

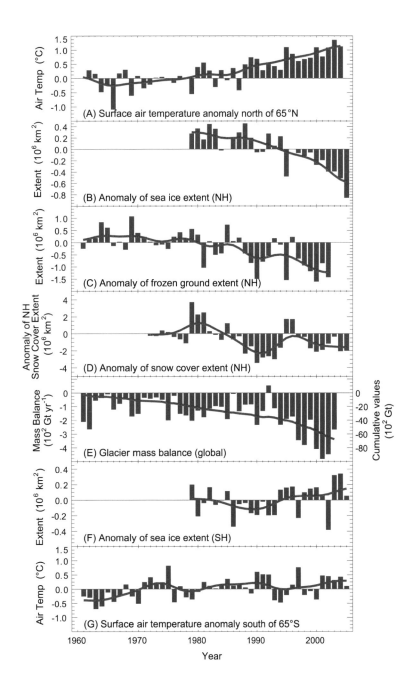

Figure 2. Departures from the long-term means of high latitude Northern and Southern Hemisphere surface temperatures (A and G) and sea ice extent (B and F), and Northern Hemisphere frozen ground extent (C) and snow cover extent (D). [IPCC Climate Change 2007, Physical Science Basis]

A number of boundary conditions and feedback mechanisms result in Polar Amplification. We will investigate feedback mechanisms at work in the Arctic where the highest latitudes are ocean environments covered by ice for much of the year (and, of course, at sea-level altitude). [This contrasts with the Antarctic where the geography is land surrounded by water, and its surface found at an average altitude of about 2500 m (8200 ft).]

Feedback: The sequence of interactions between climate controls determines how Earth's climate system responds to a disturbance (or perturbation) of the boundary conditions of the system. These interactions constitute *feedback* in the system. As defined by the IPCC, "An interaction mechanism between processes in the climate system is called a **climate feedback** when the result of an initial process triggers changes in a second process that in turn influences the initial one. A positive feedback intensifies the original process, and a negative feedback reduces it." There are many feedback mechanisms or processes in the climate system.

A feedback that reinforces the original process is called a *positive feedback*. An example of a positive climate feedback is the water vapor – greenhouse effect. An increase in surface temperature enhances evaporation at Earth's surface which introduces more water vapor into the atmosphere. Being a strong absorber of IR, greater concentrations of water vapor trap more terrestrial radiation which results in heating and a further increase in atmospheric temperature.

A feedback that tends to reduce the process that caused it is called a negative feedback. An example of negative feedback is the interaction of plant growth with atmospheric CO_2 concentration. Increasing CO_2 concentration due to human activity, or other causes, stimulates plant growth. Greater photosynthesis requires additional CO_2, and the source of this CO_2 is the air. This removal of CO_2 from the atmosphere acts to reduce atmospheric CO_2 concentrations, thereby slowing the rate of increase of CO_2.

In summary, a feedback which strengthens the original forcing mechanism is a positive feedback. A feedback which weakens the original mechanism is a negative feedback.

Incident solar radiation is a primary boundary condition of Earth's climate system. The albedo of Earth's surface determines the fraction of incident solar radiation that is converted to heat. (The lower the albedo, the greater the percentage of incident radiation absorbed and subsequently radiated upward as heat.) Therefore, all other factors being equal, the air temperature over a high albedo surface (e.g., ice cover) is lower than over a low albedo surface (e.g., open water).

5. The albedo of an ocean water surface is considerably lower than the albedo of sea ice, that is, a water surface absorbs more of the sunlight striking it than does an ice surface. (Refer to your textbook to review albedo values over different surfaces.) Hence, shrinkage of the Arctic sea ice cover and the attendant increase in ice-free ocean surfaces would result in [(*__greater__*)(*__about the same__*)(*__less__*)] absorption of solar radiation during the Arctic summer.

6. This absorption of solar radiation produces [(*__higher__*)(*__minimal change in__*)(*__lower__*)] sea surface temperatures.

7. With higher air and water temperatures in the Arctic, we would expect the rate of melting of Arctic sea ice cover to increase and lead to further warming. This is an example of [(*positive*)(*negative*)] feedback. This ice/albedo feedback is summarized graphically in **Figure 3**.

8. In Figure 3, "Thinner Ice Cover" and "Less Snow Cover" cause "More SW (solar short wave radiation) to be absorbed in system" because thinner ice cover and less snow cover [(*increases*)(*decreases*)] the surface albedo.

9. In Figure 3, evaporation is accounted for in the box entitled "Increased Summer Open Water". Less sea-ice cover and higher sea surface temperatures in the Arctic Ocean would mean [(*higher*)(*lower*)] rates of evaporation and a more humid lower atmosphere.

10. Another sequence of processes involves feedback related to clouds. A more humid atmosphere in the Arctic would likely increase the cloud cover. Clouds cause both cooling (by reflecting sunlight to space) and warming (by absorbing outgoing infrared radiation from Earth's surface and radiating some of that energy downward). During the long dark polar winter, clouds would have a warming effect. Since the primary process at this time of the year is radiative cooling, this wintertime situation is an example of a [(*positive*)(*negative*)] feedback on temperature.

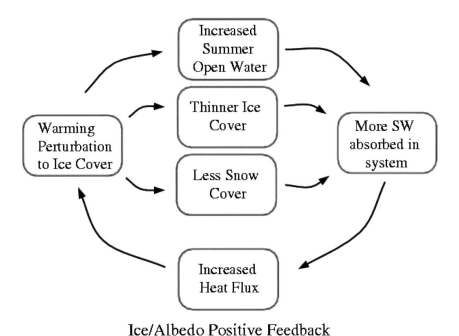

Ice/Albedo Positive Feedback

Figure 3.
A graphical representation of the ice/albedo positive feedback operating in the Arctic Ocean. Boxes enclose changes in specific climate controls. "SW" refers to solar short wave radiation. Arrows represent interactions between controls directed from a cause to an effect. (From Marika Holland, 2000. "Variability in Arctic Sea Ice: Causes and Effects." UCAR. *http://www.asp.ucar.edu/colloquium /2000/Lectures/holland.html*).

11. In summer, the impact of greater cloud cover depends on the height of the clouds. Cooling would prevail with an increase in low cover, thereby providing [(*positive*)(*negative*)] feedback. Warming would likely accompany an increase in high cloud cover, thereby providing the opposite feedback.

12. The presence or absence of snow cover has significant impacts on weather and climate. Because of higher temperatures in recent decades the extent of seasonal snow cover has decreased. In Figure 3, follow the path from "Less Snow Cover" through short wave radiation absorption and onward. This reduction in snow cover is a [(*positive*)(*negative*)] feedback mechanism, further strengthening Polar Amplification.

Snow Cover: In winter, the impact of snowstorms on weather and climate lingers long after the storm ends. **Figure 4** is a visible MODIS image of the Great Lakes area showing bare ground to the south and southwest, snow cover around Lake Superior extending southward, open water in Lakes Superior and Michigan, frozen lakes in the northwest, and clouds in the northeastern portion of the image. Note the open-water darker lake surfaces, a demonstration of water's lower albedo. Compare the albedo of the unfrozen Great Lakes with the albedo of the frozen Minnesota lakes in the northwestern part of the image.

Figure 4.
Great Lakes area with snow cover, bare ground, open water, and lake ice, 24 December 2011. [MODIS]

13. In Figure 4 it can be seen that the snow cover is highly reflective of sunlight, returning large amounts of incident solar energy back to space. The lower-albedo bare ground reflects less of the sunlight striking it. Assuming that in the cloud-free regions in the image it was midday and the rate of incoming solar radiation was relatively uniform over the area, the [(*snow-covered*)(*bare*)] ground would experience a greater heating effect by the sunlight.

14. Imagine the same snow-covered and bare areas were at midday suddenly covered by a highly-reflective cloud or aerosol (e.g., smoke, dust) layer as in the northeastern portion of the image. With such a transformation, the [(*snow-covered*)(*bare*)] ground would experience the greater change in the sunlight's heating effect. This example is intended to demonstrate only one of numerous changes in environmental conditions that can lead to complex adjustments in Earth's climate system.

Summary: The presence or absence of snow and ice cover controls patterns of heating and cooling over Earth's surface more than any other surface feature. The higher latitudes of the Northern Hemisphere are especially impacted, resulting in Polar Amplification. Polar Amplification, in turn, impacts other boundary conditions of Earth's climate system.

Investigation 12A:

CLIMATE CHANGE AND RADIATIVE FORCING

Driving Questions: *What is climate change and how can it be determined? What is radiative forcing and what makes it important to climate science?*

Educational Outcomes: To define climate change and *radiative forcing*. To identify factors that affect climate, the agents and mechanisms which exert forcings that alter climate, the relative magnitude of each factor, and the evaluation of the total radiative forcing from the group of factors.

Objectives: After completing this investigation, you should be able to:

- Describe climate change and how it is objectively determined.
- Explain the IPCC concept of radiative forcing as a way to commonly describe the influence of various factors that can cause climate change.
- List and explain the principal causes of climate change, including their relative impacts leading to warming or cooling.

Climate Change

The scientific consensus arrived at by the Intergovernmental Panel on Climate Change (IPCC) is that climate change is very likely occurring as it is "unequivocal" that global Earth is warming. Here we consider how we can quantify the changes in boundary conditions that bring about this temperature increase.

As stated in the first week of this course, **our global climate is fundamentally the story of solar energy received by Earth being absorbed, deflected, stored, transformed, put to work, and eventually emitted back to space.** Climate describes the slowly varying aspects of the atmosphere-hydrosphere-land surface system. It encompasses and bounds the broad array of weather conditions and impacts that arise from the complex interplay of the sub-systems of Earth's climate system in response to this energy flow.

The relative amounts of incoming and outgoing energy, on a global basis, determine whether or not Earth is in a steady-state condition, cooling, or warming. Earth's climate system, essentially an open physical system regarding energy, is sustained by a continuous supply and removal of energy. It achieves a steady-state condition when its properties (e.g., temperature) do not vary in statistically significant ways when averaged over time. **Climate change refers to a change in the state of the climate that can be identified (e.g., using statistical tests) by changes in the mean and/or the variability of its properties, that persists for an extended period, typically decades or longer.** Climate change results from natural causes and/or human (anthropogenic) activities.

Radiative Forcing

To give focus to climate change, the IPCC in its First Assessment Report (1990) emphasized a new concept, ***radiative forcing***. Radiative forcing refers to the change in the net vertical radiation flow (expressed in W/m^2) in the climate system at the top of Earth's troposphere that is caused by the addition of greenhouse gases (e.g., CO_2) or other changes (including solar radiation, Earth's surface albedo, and aerosols). The IPCC describes radiative forcing as a modeling concept that is a simple but important means of estimating and comparing the impacts of different natural and anthropogenic radiative causes on the surface-troposphere climate system.

Radiative forcing is the change in net (difference between downward solar and upward longwave) radiant energy flow at the upper limit of the planet's atmosphere, assumed on Earth to be at the tropopause, under a specified set of conditions. When the net flow of energy is into the planetary climate system, it produces positive radiative forcing; when the net flow is to space, the radiative forcing is negative. In the case of Earth, radiative forcing values are typically changes measured relative to the mean state of Earth's climate system at the start of the industrial era (about 1750).

Follow the Energy! The essentials of radiative forcing are embodied in various AMS CEM scenarios, included those first presented in *Investigation 1B*'s **Follow the Energy!** Go to the course website and click on "AMS Conceptual Energy Model".

1. Set the AMS CEM at One Atmosphere, Sun's Energy: 50%, 200 Cycles, and Introductory Mode. Run the model. It shows that the mean energy content in the planet's climate system was [(***0.5***)(***2.3***)(***4.7***)(***10.7***)] energy units.

2. Now change only the model's Sun's Energy setting from 50% to 100%. This is analogous to doubling incident solar radiation. After running the model at this higher rate of solar irradiance, it is seen that the mean energy content in the planet's climate system changed to [(***0.5***)(***2.3***)(***4.7***)(***10.7***)] energy units.

3. Comparing the two runs of the model shows that increasing the amount of solar energy entering the planet's climate system would produce positive radiative forcing and attendant warming. Change in the rate at which the Sun delivers energy to the planet is an example of a(n) [(***natural***)(***anthropogenic***)] climate change.

4. Now compare the two runs of the model in terms of its response if the rate at which the Sun delivers energy to the planet had been the opposite, that is, it decreased from 100% to 50%. The comparison shows that the reduced incident solar radiation would result in [(***positive***)(***negative***)] radiative forcing of the planet's climate system, with accompanying global cooling.

Adding an Atmospheric Layer: Switching the Atmospheres setting from One Atmosphere to Two Atmospheres subjects energy units passing through the atmospheric component of the model to a another application of the model's **Rule 2**, which states, "During each cycle, any energy unit in the atmosphere will have an equal chance of moving downward or upward." A rising energy unit from the planet's surface will first reside in the first (lower) atmosphere. In the next cycle of play, it will move upward to the second (higher) atmosphere or downward to the planet's surface. Once in the second (higher) atmosphere, it has during the subsequent cycle equal chances of escaping upward to space or moving downward to the first (lower) atmosphere.

5. Set the AMS CEM at Two Atmospheres, Sun's Energy: 100%, 200 Cycles, and Introductory Mode. Run the model. After running the model with the two-atmosphere configuration, the mean energy content in the planet's climate system is [(*0.5*)(*2.3*)(*4.7*)(*10.7*)] energy units.

6. Compare the mean energy content in the planet's climate system from Item 2, which was set at One Atmosphere, Sun's Energy: 100%, 200 Cycles, and Introductory Mode, with the two-atmosphere value from Item 5. The comparison shows that adding a second atmosphere [(*increases*)(*decreases*)] the energy content of the planet's climate system.

7. The change from One Atmosphere to Two Atmospheres in the model settings results in [(*positive*)(*negative*)] radiative forcing of the planet's climate system.

8. The change from the One Atmosphere setting to Two Atmospheres in the AMS CEM is analogous to a doubling of heat-absorbing greenhouse gases in Earth's atmosphere. Comparison of the amounts of energy residing in the planet's climate system before and after the doubling shows that such an enhanced greenhouse effect is an example of radiative forcing that [(*increases*)(*decreases*)] the energy content of the planet's climate system.

9. This change in the energy content of the planet's climate system would result in a [(*colder*)(*constant*)(*warmer*)] mean global temperature.

10. The addition of greenhouse gases to Earth's atmosphere is an example of climate change largely due to [(*natural*)(*anthropogenic*)] causes.

Comparing Principal Radiative Forcing Components: The strength of the concept of radiative forcing is that the influence of different factors that cause climate change can be quantified using common units (W/m^2), which allows them to be compared and tracked. This is an essential prerequisite to better understanding the agents and forcing mechanisms that can produce climate change, and to scientifically predict climate change beyond simply extrapolating observed current climate trends into the future.

Figure 1 summarizes the principal components and impacts of the radiative forcing of climate during the industrial era between 1750 and 2005. Their forcing is reported in W/m^2, with positive forcings leading to warming of climate and negative forcings leading to cooling. The dots at the end of the thick bars indicate the best estimates of the values and the

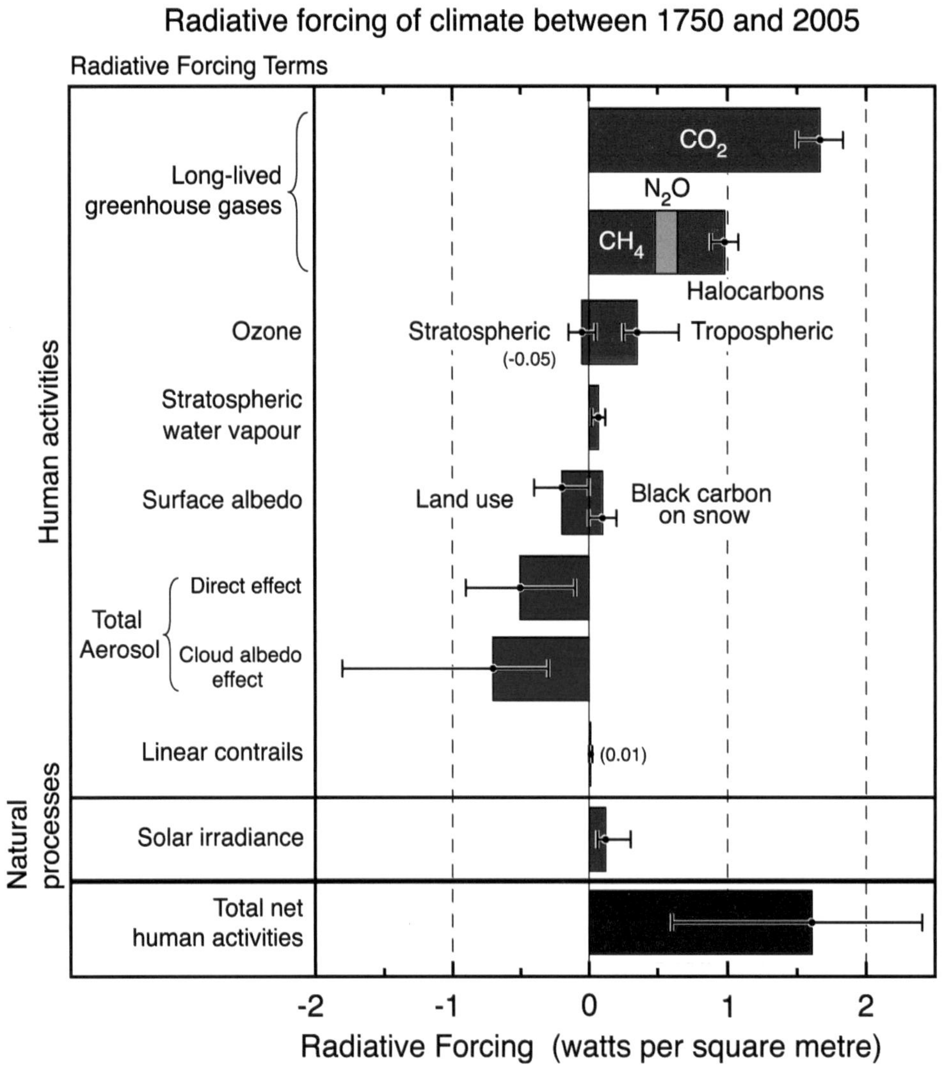

Radiative forcing of climate between 1750 and 2005

Figure 1.
Summary of the principal components of the radiative forcing of climate between 1750 and 2005. Solar irradiance refers to incident solar radiation. [IPCC *Climate Change 2007, The Physical Science Basis*, FAQ 2.1, Figure 2]

thin horizontal bars passing through the dots represent the judged possible variation of the particular value, referred to as a "range of uncertainty."

11. According to the list of radiative forcing terms along the left in Figure 1, the only natural process that produced radiative forcing during the industrial era is [(*stratospheric water vapor*)(*surface albedo*)(*solar irradiance*)]. The figure shows the forcing is positive; therefore, it has a warming impact. [Note: According to the IPCC, this particular forcing component has increased slightly during the industrial era.]

12. Violent volcanic eruptions (discussed in Investigation 11A), which occur infrequently, are not listed in Figure 1 as radiative forcing agents. This is because

[(***they cannot cause global-scale temperature changes***)(***the global temperature changes they can cause are typically observable only for a year or two***)].

13. According to Figure 1, the major positive radiative forcing (warming) arises from anthropogenically-produced [(***albedo changes due to land use***)(***total aerosol effects***)(***carbon dioxide***)].

14. According to Figure 1, the total radiative forcing due to human activities during the industrial era is equivalent to a positive (warming) [(***0.6***)(***1.0***)(***1.6***)(***2.4***)] W/m^2. This value is within a range of uncertainty between about 0.6 and 2.4 W/m^2 as shown by a horizontal line in the graph. This range of uncertainty concerning the value is a major reason why the global temperature increases predicted for the 21st century under different scenarios include the best estimate of temperature and likely range.

15. Given the value for total radiative forcing from the previous question, scientists would predict that the combined anthropogenic and natural perturbations present today would result in a troposphere today and into the future that is [(***warmer than***)(***the same as***)(***cooler than***)] the troposphere of the year 1750. And they would be able to make a prediction regarding the range of future tropospheric temperatures based on the magnitude of the total radiative forcing value.

[Note: How do these radiative forcings compare with the average solar energy intercepted by Earth? The value of average incident solar radiation at Earth's surface is determined as follows. The total solar energy intercepted by Earth is equal to the solar constant (1368 W/m^2) times the cross-sectional area of the planet. Because Earth's total surface area is four times its cross-sectional area (area of circle = πr^2, area of sphere = $4\pi r^2$), the average solar energy received per square meter over the entire surface of the globe is one-quarter the solar constant (i.e., 342 W/m^2).]

Summary: Climate scientists are demonstrating considerable progress in identifying the factors that affect climate, the agents and mechanisms which exert forcings that alter climate, the relative magnitude of each factor, and the evaluation of the total radiative forcing from the group of factors.

This consideration of the Earth environment as a system is powerful in its potential to better understand climate and climate change. The highly complementary empirical and dynamic approaches to climate study reveal the workings of Earth's climate system. As stated at the beginning of this course, the empirical approach allows us to construct <u>descriptions</u> of climate, and the dynamic approach enables us to seek <u>explanations</u> for climate. Each has powerful applications. In combination, the two approaches enable us to explain, model, and predict climate and climate change.

Acknowledgments: Dr. James F. Lubner (University of Wisconsin – Milwaukee) provided constructive analysis concerning radiative forcing as depicted in this investigation. Thanks are extended to Drs. David W. Fahey and Ellsworth G. Dutton (Earth System Research Laboratory/NOAA) for their helpful suggestions in Prof. Lubner's analysis.

THE OCEAN IN EARTH'S CLIMATE SYSTEM

Driving Questions: *How important is the ocean in the climate system and climate change? How are oceanic conditions changing?*

Educational Outcomes: To describe why the ocean is the major component of Earth's climate system in terms of mass and energy. To show the significant increase in the energy content in Earth's climate system, particularly in the ocean, in recent decades that evidence climate change. To explain what changes are occurring in the ocean and their observed impacts on Earth's climate system.

Objectives: After completing this investigation, you should be able to:

- Explain why the ocean is the dominant energy and mass reservoir and most influential component of Earth's climate system.
- Describe what influences the ocean has on and receives from the rest of the climate system.
- Explain what changes are occurring in the ocean and what the consequences may be for Earth's climate system.

The Ocean Component of Earth's Climate System

Earth's climate system, which has been the subject of this course, encompasses the "spheres" of the atmosphere, hydrosphere (including the ocean), cryosphere, geosphere, and biosphere along with their interactions and the variability and changes within and because of boundary conditions. Traditional climate studies have often been restricted to conditions at Earth's surface primarily due to the workings of the atmosphere. We are culminating our primary investigations of Earth's climate system by focusing on the ocean, the system's major component in terms of mass and energy.

While we live on the land and interact most with the atmosphere, the major component of Earth's climate system in terms of mass and energy is the ocean. The ocean covers about 71% of Earth's surface and makes up 99.8% of the mass of the fluid portions of Earth's system (air and water). The high specific heat of water combined with the size of the ocean reservoir gives the ocean exceptional capacity to store and transport heat. Therefore, just from the mass and energy characteristics, the variability of the fluid portions of the planet implies that the ocean merits major consideration when seeking understanding of Earth's climate system and its changes.

Change in Heat Content of Climate System Reservoirs: **Figure 1** shows the estimated heat energy content changes in various parts of the climate system for two periods: 1961-2003 (blue bars) and 1993-2003 (red bars). All of these values represent increases in the heat storage of those portions of the climate system during the respective time periods. The heat units are in 10^{22} J. [Note: 1 joule (J) = 0.239 calorie]

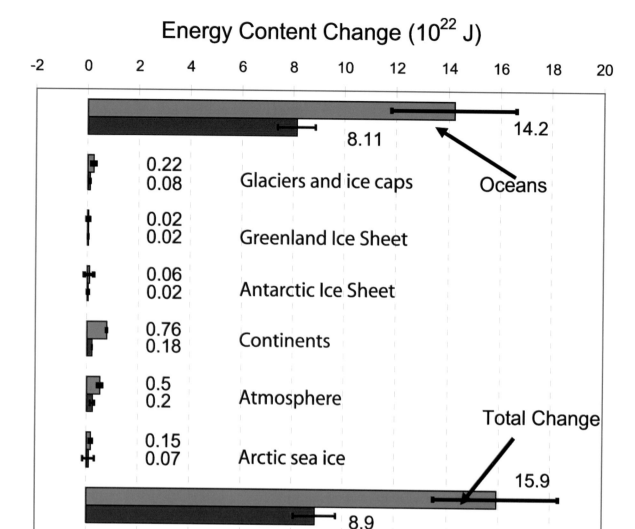

Figure 1.
Energy content change of portions of the climate system, from 1961-2003 (blue) and from 1993-2003 (red). (IPCC)

1. For all of the energy content changes shown in Figure 1, the values varied considerably in magnitude. However, the estimated values were all positive. What does the positive energy change from 1961 to 2003 imply about these parts of the climate system? [(*all cooled*)(*some cooled and some warmed*)(*all warmed*)]

2. The total change in the energy content of the components of Earth's climate system shown in Figure 1 from 1961 to 2003 was [(*14.2*)(*15.9*)(*24.8*)] x 10^{22} J.

3. Figure 1 indicates that in the 1993-2003 time period a total of 8.9 x10^{22} J were added to Earth's climate system's heat content. Therefore, the climate system's heat content must have increased [(*7.0*)(*8.11*)(*15.0*)] x 10^{22} J during the earlier time period from 1961 to 1993.

4. Comparison of the 1961-1993 total change with the 1993-2003 total change implies that the annual rate at which heat energy was being added to Earth's climate system was generally [(*decreasing*)(*steady*)(***increasing***)] from year to year.

5. Compare the numbers for the energy content changes in the continents and the atmosphere over the period 1993-2003 (red bars). For these two portions of the climate system over this latter period, the energy content change(s) of [(***continents were much greater***) (*continents and atmosphere were about equal*)(*the atmosphere was much greater*)].

6. Combining the values for the ice components of the system (glaciers, ice sheets and sea ice) provides a total energy content change of 0.45×10^{22} J during the 1961-2003 time period. Compared to the energy content change for the continents and atmosphere during the same time period, the total ice energy content change was nearly equal to the change within the [(***atmosphere***)(*continents*)] component.

7. Compared with the energy content change for oceans for either period, the changes of all the other portions of the climate system <u>combined</u> were [(***much smaller***)(*about the same*) (*much greater*)].

8. Using the value for the ocean change over the 1993-2003 period compared to the system's total change for the same 1993-2003 time period (red bars), the ocean accounted for about [(*10%*)(*25%*)(*50%*)(***75%***)(*90%*)] of the total energy content increase in the system. This demonstrates the prominent role the ocean must play in determining climate and climate change.

Global Warming Impacts on the Ocean: The *IPCC AR4, Climate Change 2007: Physical Science Basis, Technical Summary*, p. 47, states, "Warming is widespread over the upper 700 m of the global ocean." Further, "The world ocean has warmed since 1955, accounting over this period for more than 80% of the changes in the energy content of the Earth's climate system." This ocean warming has several critical impacts on the climate system.

9. The temperature change of ocean water leads to changes in evaporation and density. Evaporation from the ocean drives the global water cycle. Warmer water has more energy available for evaporation. In locations where precipitation is less than evaporation, we would expect that excess evaporation would lead to surface ocean waters that become [(*less*)(***more***)] saline (as salts are left behind in the evaporation process).

"There is now widespread evidence for changes in ocean salinity at gyre and basin scales in the past half century with the near-surface waters in the more evaporative regions increasing in salinity in almost all ocean basins. These changes in salinity imply changes in the hydrological cycle over the oceans." (IPCC AR4, *Physical Science Basis, Technical Summary*, p.48)

10. As seawater warms, it becomes less dense. Less dense water means that the same mass of water would occupy a greater volume. The result of a warming ocean would be [(*rising*)(*lowering*)] sea level.

"The global average rate of sea level rise measured by TOPEX/Poseidon satellite altimetry during 1993 to 2003 is 3.1 ± 0.7 mm yr^{-1}." (IPCC AR4, *Physical Science Basis, Technical Summary*, p.49

11. Another crucial gas dissolved in the ocean is oxygen. Dissolved oxygen is essential to the oceanic biosphere. With a warmer ocean, [(*more*)(*less*)] O_2 would be dissolved in the water.

In fact, the IPCC states, "The oxygen concentration of the ventilated thermocline (about 100 to 1000 m) decreased in most ocean basins between 1970 and 1995. These changes may reflect a reduced rate of ventilation linked to upper-level warming and/or changes in biological activity." (IPCC AR4, *Physical Science Basis, Technical Summary*, p.48)

Although there is no known evidence at present of its occurrence, one fear is that the warmer, less dense water may resist sinking in the North Atlantic Ocean basin. It is that sinking that drives the global meridional overturning circulation, a massive inter-basin heat engine for the ocean system. And, warmer ocean surface waters in tropical regions could lead to greater tropical cyclone activity through an increase in the number of storms, by storms becoming more intense, or both.

Ocean Acidification: Associated with the changes already mentioned is the impact of increased carbon dioxide in the atmosphere and ocean.

Of the carbon dioxide that is emitted into the atmosphere, about 56% is taken up by dissolving in the ocean. The ocean is normally slightly alkaline. The CO_2 that dissolves in ocean waters forms a weak acid solution that increases the acidity of the waters, causing pH values to decrease (the lower the pH, the higher the acidity). One consequence is that the calcium carbonate shells or structures of many ocean creatures become more soluble in more acidic waters. Ocean life will have difficulty living in a more acidic ocean. "The uptake of anthropogenic carbon since 1750 has led to the ocean becoming more acidic, with an average decrease in surface pH of 0.1 units." (IPCC AR4, *Physical Science Basis, Technical Summary*, p.48)

Figure 2 is a summary of these interrelated effects over the global ocean. Red arrows represent heat and carbon dioxide changes, blue arrows indicate the net water mass movements (evaporation versus precipitation), black arrows signify changes in the thermocline and calcium carbonate dissolution level and yellow circles represent the decreasing pH values. Above the colored portion of the sea along with ice extent is the latitudinal rise of sea level. Much of that increase is due to the warming of seawater. The rise of sea level due to thermal expansion is referred to as *thermosteric* or *steric* sea level change.

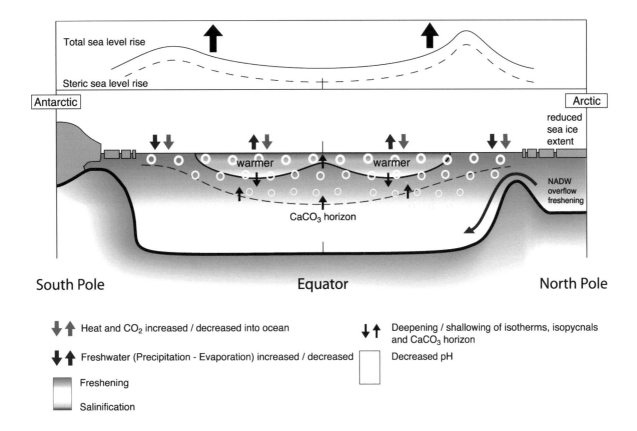

Figure 2.
Schematic of ocean changes including temperature, carbon dioxide absorption, evaporation effects. [IPCC]

12. As seen in this investigation, we [(*can*)(*cannot*)] ignore the role of the ocean if we want to more completely understand Earth's climate system and climate change.

13. The ocean plays a [(*minor*)(*major*)] role in the climate system through its cycling of mass and energy.

Summary: We now go back to reconsider the **Earth's Climate System paradigm**, which was addressed in Investigation 1A. This statement provided us with a guiding definition of climate, the climate system and the ways humans interact with and affect that system of which they are a part. *"Climate can be explained primarily in terms of the complex redistribution of heat energy and mass by Earth's coupled atmosphere/ocean system."* Many of the investigations in this course examined the boundary conditions that derived from the physical processes at work on and in the climate system. We also looked at the mean state of the system and its variability within those boundaries.

"Scientific research focusing on key climate processes, expanded monitoring, and improved modeling capabilities are already increasing our ability to predict the future climate." A major environmental monitoring effort is underway, called GEOSS - the Global Earth Observation System of Systems. Data from this effort has provided the insight for the ocean

changes we noted previously. And, it will provide the basis for understanding the changes currently occurring as well as the computer models that will inform policy decisions for future human activities in our climate system.

Your knowledge of Earth's climate system will be enhanced by considering both its empirical and dynamic perspectives. The more we know about the environment in which we live, the more informed we will be to work toward making it a better place for everyone.

13A - 1

Investigation 13A: VISUALIZING CLIMATE

Driving Question: *How are the unique climate characteristics of different locations on Earth routinely displayed so their climates can be systematically compared?*

Educational Outcomes: To describe how climate has traditionally been held to be the synthesis of weather conditions, both the average of parameters, generally temperature and precipitation, over a period of time and the extremes in weather over the period of record. To portray the statistical descriptive information of climate in graphs, typically as the magnitude of the average value (or extremes) versus the months of the year. To examine climographs of various locations which show relationships between temperature and precipitation during the yearly cycle.

Objectives: After completing this investigation, you should be able to:

- Portray the statistical climate values of mean monthly temperature and average total monthly precipitation in the graphical form called the climograph.
- Compare and contrast temperature and precipitation distributions on climographs from different locations.
- Explain how climograph patterns can be related to climate forcings.
- Relate certain patterns of temperature and precipitation to particular climate classification types.

Portraying Climates of Various Locations With Climographs

Every location on Earth has climate characteristics that distinguish it from other locations. It is desirable to systematically describe these characteristics so that the climates of various locations can be compared. This investigation focuses on climate as described by average values alone. It is important to remember that by relying on average values, only a generalized and incomplete picture of the climate is provided. A *climograph* is a commonly used tool to portray the climate of a given locale and compare climates from various places. A climograph can be drawn to show monthly mean temperatures and average precipitation totals (rain plus melted snow and ice) for a single station through the year on the same graph. **Figures 2 - 7** are climographs for six locations in the United States which give examples of major climate types discussed in the Climate Classifications of Chapter 13 in the textbook.

A climograph can provide at a glance the magnitudes and ranges of monthly mean temperatures and average monthly precipitation totals throughout the year. These statistics are genetically tied to various climate-determining factors which vary systematically from place to place. It is possible to relate distributions of temperature and precipitation to specific climatic forcings to gain a more comprehensive understanding of the controls of the climate in a specific area. Because these controls and the climates which result have significant impact on other aspects of the Earth system (human housing, fresh water supply, agriculture, health, etc.), such an understanding has widespread applications. It is also desirable to have a shorthand classification for the major types of recurrent temperature and precipitation patterns so one may be able to generalize about climates up to the global scale.

By convention, as shown in **Figures 2 - 7**, climographs are usually constructed with time of year displayed horizontally across the base of the graph. The initial letter of the month is listed at mid-month along the bottom, with the precipitation scale along the left vertical axis and mean temperature scale on the right axis. Mean monthly precipitation totals (rain plus melted snow and ice) are presented as a bar graph. The mean temperature values are plotted as points and connected by a curve. In the U.S., climate data are prepared with precipitation in inches and temperatures in degrees Fahrenheit. **Use the data and grid in <u>Figure 1</u> below to make a climograph for Boston, MA. Mark a short horizontal line at mid-month to note the position of the mean total precipitation value of that month and fill in the space below to create a bar. (Use Figures 2 - 7 as a guide.) Place a dot at mid-month at the level denoting the mean monthly temperature value. When all the months are plotted, connect the temperature dots with curved line segments to represent the march of average monthly temperature.**

1. Your completed climograph for Boston, MA (Figure 1) shows that the mean monthly temperature rises from near freezing during the winter months (Dec, Jan, and Feb) to means around 70 °F during the summer months (June, Jul, and Aug) and then falls as winter approaches. The observed temperatures from which the means are computed result mainly from the seasonal swing of solar heating, which in turn is largely determined by latitude. As a general rule, the higher the latitude the lower the winter season temperatures. The lowest mean monthly temperature in Boston occurs in [(*<u>January</u>*) (*<u>December</u>*)].

2. This minimum monthly temperature [(*<u>is</u>*)(*<u>is not</u>*)] within a month or so of the time of minimum receipt of solar radiation in a mid latitude, Northern Hemisphere location.

3. Where solar heating varies significantly between the winter and summer solstices, the range of temperature, indicated by the amplitude of the temperature curve on a climograph, is relatively large. Where the amplitude is relatively small, the seasonal temperature contrast is also small. Examine the temperature curve on the climograph for Hilo, HI (**Figure 2**). The range of mean monthly temperatures for Hilo is about [(*<u>5</u>*)(*<u>20</u>*) (*<u>30</u>*)] Fahrenheit degrees.

4. From the shape of the curve and range of temperature, it is evident that Hilo experiences relatively [(*<u>little</u>*)(*<u>significant</u>*)] variation in solar heating through the course of a year.

5. The highest mean monthly temperature in Hilo occurs in August. This temperature is about [(*<u>76</u>*)(*<u>86</u>*)] °F.

6. The lowest mean monthly temperature is about [(*<u>71</u>*)(*<u>81</u>*)] °F in both Jan. and Feb.

7. These temperatures suggest that Hilo is a [(*<u>high</u>*)(*<u>low</u>*)] latitude location.

Boston (42 N, 71 W)

Month	Temp.(F)	Precip.(in)
J	29.3	3.92
F	31.5	3.30
M	38.9	3.85
A	48.3	3.60
M	58.5	3.24
J	68.0	3.22
J	73.9	3.06
A	72.3	3.37
S	64.7	3.47
O	54.1	3.79
N	44.9	3.98
D	34.8	3.73

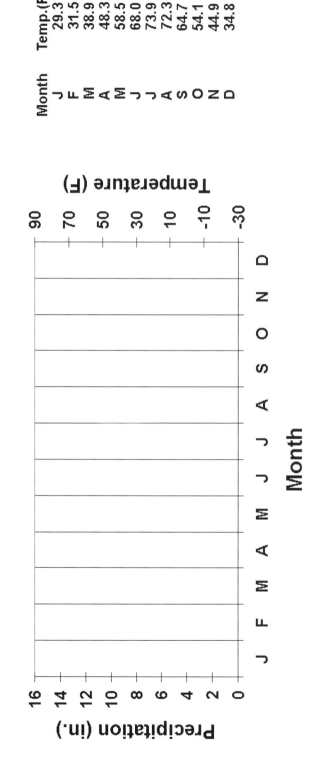

Figure 1.
Humid Continental Climate (Dfa) - Boston

8. The month of occurrence of the highest mean temperature suggests that Hilo is located in the [(***Northern***)(***Southern***)] Hemisphere.

9. Temperature and temperature range can also be influenced by large bodies of water (ocean or large lake). Generally speaking, a maritime influence will moderate temperatures in places that would normally be colder in winter and warmer in summer based on latitude alone. The seasonal range in temperatures is likely to be less due to a maritime influence; that is, the temperature curve on the climograph will exhibit less amplitude. In continental locations or locations downwind of large land masses, temperatures tend to be higher in summer and lower in winter. As a result, the seasonal temperature range will be far greater than for places surrounded by or downwind of a large water body. It is likely that Hilo's relatively low annual temperature range [(***is***)(***is not***)] moderated by the surrounding Pacific Ocean.

10. Examine the climograph for Fairbanks, AK (**Figure 6**). The highest mean monthly temperature is [(***62***)(***82***)] °F.

11. The lowest mean monthly temperature for Fairbanks is [(***–10***)(***10***)] °F.

12. These temperatures suggest that Fairbanks is a [(***high***)(***low***)] latitude location.

13. The average monthly temperatures in Fairbanks cover a range of about [(***70***)(***50***)(***30***)] Fahrenheit degrees.

14. This range of temperatures suggests that Fairbanks has a [(***continental***)(***maritime***)] climate.

15. Compare the annual temperature range for Boston, on the Atlantic coast, (**Figure 1**) and Seattle, WA, near the Pacific coast, (**Figure 5**). Seattle's annual temperature range is [(***greater than***)(***less than***)] that of Boston.

16. These cities are at approximately the same latitude and both are located near the coast. The climate control causing the difference in annual temperature range is the influence of the prevailing westerly wind at both locations. For Seattle, the temperature range is influenced mainly by the [(***ocean***)(***continent***)] which is upwind and in Boston by prevailing winds blowing from the continent.

17. Climate classification systems allow climate differences and similarities to be expressed in a "shorthand" form. The broad-scale climate boundaries in the *Köppen* climate classification system (see Chapter 13 of the text) are based on patterns in annual and monthly mean temperature and precipitation, which closely correspond to the limits of vegetative communities. The major classifications of *Tropical Humid* (A), *Subtropical* (C), *Snow Forest* (D), and *Polar* (E) are based on temperature; the group *Dry* (B) is based on precipitation; and the group *Highland* (H) applies to mountainous regions. Temperatures for both Fairbanks and Boston place them both in the [(***Tropical Humid (A)***)(***Subtropical (C)***)(***Snow Forest (D)***)(***Polar (E)***)] climate classification.

18. The second letter of the Boston (Figure 1) and Fairbanks (Figure 6) classifications corresponds to seasonal precipitation regimes with an "f" signifying year-round precipitation. According to their climographs and climate classifications, Boston and Fairbanks have [(*similar*)(*very different*)] month-to-month uniformity in their precipitation regimes.

19. Arid and Semiarid climates can be caused by several climate controls. Locations on the eastern side of planetary-scale, semi-permanent subtropical high pressure systems, such as those occurring around 30 degrees N in the Atlantic and Pacific Ocean basins, are characterized by subsiding air, which inhibits cloud formation and precipitation. The western side of such systems, by contrast, tends to be humid. The cause of dryness in Tucson, AZ, for example, is due mainly to its position [(*east*)(*west*)] of a subtropical high pressure system which persists off the southwest U.S. coast in the Pacific.

20. Atlanta, GA, at about the same latitude as Tucson, but in the southeastern United States, is humid because it is located [(*east*)(*west*)] of the Bermuda-Azores subtropical high over the Atlantic Ocean basin.

21. Dry or wet conditions can also be caused by location upwind or downwind of a mountain range. Areas to the lee of high mountains tend to be dry because of the "wringing out" of moisture on the wet, windward slopes (due to orographic lifting, cooling and condensation) and the compressional warming of air which occurs as the air descends on the leeward slopes. The atmospheric stability caused by cold ocean currents offshore can also prevent precipitation by stabilizing the overlying air and inhibiting convection. On the other hand, instability cloudiness and precipitation can occur if ocean currents are relatively warm. Tucson, located downwind of the Coastal Ranges and the cold California Current, is [(*probably*)(*not likely*)] drier because of the influence of mountains and ocean currents.

22. The **Figure 8** world map showing the Koeppen (or Köppen) climate classification system demonstrates the actual application of climatic controls. For example, on this Mercator projection map, horizontal lines (if they were drawn) would represent constant latitudes. Therefore, at similar latitudes across the Eurasian land mass, Europe to the west is shown in purple, indicating a temperate climate while eastern Asia is in yellow indicating a cold type climate. The climate control primarily at work in these climate types is the prevailing wind circulation in relation to [(*elevation*)(*proximity to large bodies of water*)(*Earth's surface characteristics*)].

23. The region of Tibet in south central Asia is shown in the greenish-brown of a polar type climate although it is surrounded by dry or temperate climates. This classification is most likely the result of Tibet's [(*elevation*)(*proximity to large bodies of water*)(*Earth's surface characteristics*)].

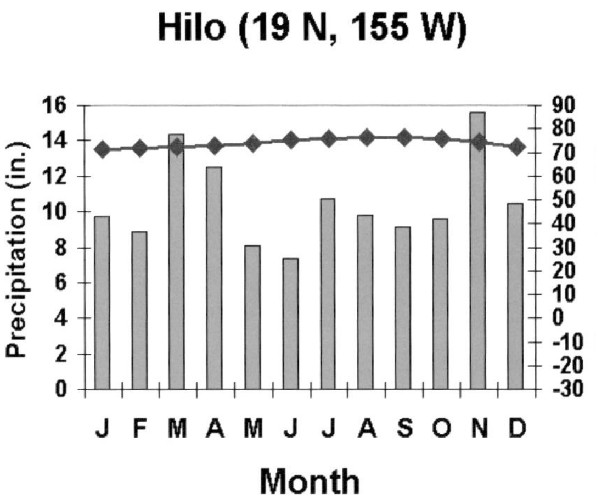

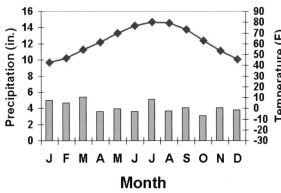

Figure 2.
Tropical Wet Climate (Af) - Hilo.

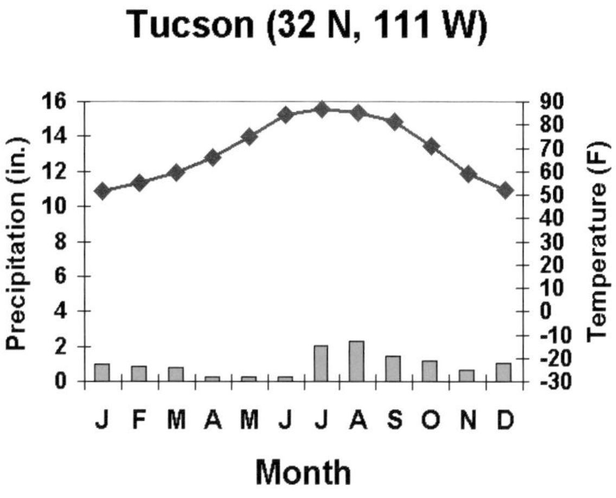

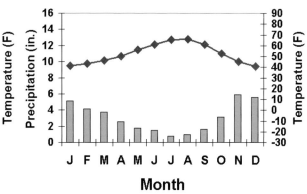

Figure 3.
Subtropical Desert Climate (BWh) - Tucson.

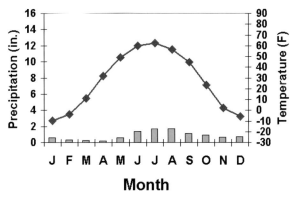

Figure 4.
Subtropical Humid Climate (Cfa) - Atlanta.

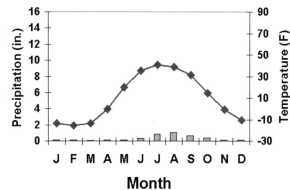

Figure 5.
Marine West Coast Climate (Cfb) - Seattle.

Figure 6.
Subarctic Climate (Dfc) - Fairbanks.

Figure 7.
Polar Tundra Climate (Et) - Barrow.

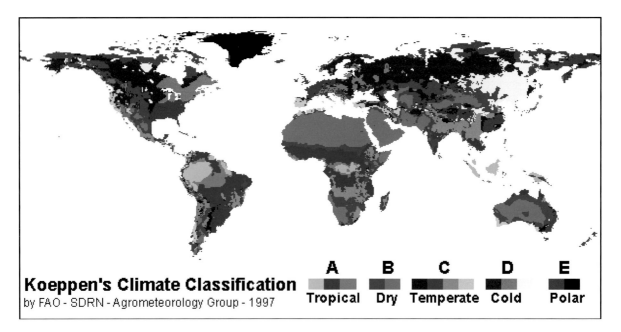

Figure 8. Koeppen's Climate classification. (*Adapted from UN Food and Agriculture Organization, Sustainable Development website.*)

Summary: Climate classification systems have been devised to simplify and organize climate types by grouping together climates having common characteristics. Such systems allow climate differences and similarities to be expressed. The climograph is a commonly used tool that shows the relationships between temperature and precipitation during the yearly cycle.

Suggestions for further investigations: For interactive maps where cities in the United States may be selected to display their climographs, go to: *http://www.drought.unl.edu/ DroughtBasics/WhatisClimatology/ClimographsforSelectedUSCities.aspx*.
For interactive maps where climographs for international cities can be selected, go to: *http://www.drought.unl.edu/DroughtBasics/WhatisClimatology/ ClimographsforSelectedInternationalCities.aspx*.

You can make your own climographs with monthly average temperatures and precipitation totals from *http://www.worldclimate.com*. Monthly and annual values are provided in both English and metric units. By inputting data to spreadsheet software, graphing can be easily accomplished allowing comparisons among stations.

Investigation 13B:

CLIMATE VARIABILITY AND SHORT-TERM FORECASTING

Driving Questions: *How are NOAA's Climate Prediction Center's temperature and precipitation outlooks interpreted? Can these outlooks show evidence of forcings of weather systems by recognized patterns of variability in Earth's climate such as El Niño/La Niña?*

Educational Outcomes: To describe how descriptive climate data can provide past and present perspectives for planning purposes. To identify sources of climatic outlooks for periods of months to a year or more. To explain how to interpret climatic outlook products.

Objectives: After completing this investigation, you should be able to:

- Identify where to find climatic outlooks for periods of months to a year or more.
- Explain how to interpret climatic outlook products.

Descriptive Climate Data, Its Uses, and Sources

Descriptive climate data provide valuable past and present perspectives for planning purposes. Farmers use their knowledge of climate to determine what crops to plant and when to plant and harvest. Utilities use climate data for planning production and distribution of energy supplies and the reallocation among types of such supplies. The building industry is interested in the design of structures, including their necessary strength, heating and cooling energy requirements, and the associated building codes that regulate them. People look to climate data as they plan future activities (e.g., outdoor weddings, vacations, and sporting events). Even more valuable would be a reliable expectation of what future climatic conditions would be.

In the U.S., NOAA's National Climatic Data Center in Asheville, NC, compiles U.S. data as well as being a depository for much worldwide data on weather and the environment. NOAA's Climate Prediction Center (CPC) and other organizations apply these data to numerical models and objective analyses in the production of near-future outlooks for climatic conditions. The term *outlook* is preferred to *forecast* because the future values are typically given in broad categories (i.e. above, below or normal) with probabilistic likelihoods of occurrence.

The CPC, one of the National Centers for Environmental Prediction of the National Weather Service, issues outlooks for temperature and precipitation for periods of six to ten days out to the year ahead. The CPC's website (*http://www.cpc.noaa.gov/*) also monitors several recognized patterns of variability in Earth's climate system such as persistent drought patterns, El Niño/La Niña phases, and climatic oscillations (North Atlantic, Arctic, Antarctic, Pacific-North American).

Figure 1 is composed of samples of the three-month outlooks for temperature (left) and precipitation (right) for the United States. [The latest three-month outlook can be seen at: *http://www.cpc.ncep.noaa.gov/products/predictions/90day.*] The sample outlooks in Figure 1 were issued on 21 April 2011 for the May/June/July 2011 time period. In the left temperature map, shades of brown/red depict regions that were expected to be above normal (A) while shades of blue depict regions expected to be below normal (B). White areas (EC) in the U.S., including Alaska, represent locations where there was no clear climatic signal and there were therefore equal chances of the seasonal averages being either above or below normal. Within each shaded area, the increasing confidence of that outcome is shown by the hue and dashed boundary lines. For example, the temperatures for the three months across much of the Southwest including Arizona, New Mexico and southwestern Texas were expected to be above normal with 50% or more likelihood. The precipitation outlook in the map to the right depicted expected precipitation totals to be equally likely as above or below normal in white while those below were in brown and above in green.

1. For the left temperature map, it was expected that the average May/June/July 2011 temperatures would be above the climatological norm in the Southwest and Gulf Coast states. With the exception of temperatures in the Northeast expected to be normal, below normal temperatures were expected generally in the [(*eastern*)(*northern*)(*southwestern*) (*southern*)] tier of states into the Mid-Atlantic region.

2. As shown in the precipitation outlook map on the right in Figure 1, above normal precipitation totals were anticipated to occur from New England to the Mid-Atlantic, in the central Midwest and along the northern Plains at the Canadian border. Generally below normal precipitation totals were expected in which area?
[(*along the western Gulf Coast*)(*Florida*)(*Northwest states*)(*Alaska*)].

3. Consider the combination of temperature and precipitation anomalies (departures from normal) anticipated for the state of Texas for the three-month period. One can characterize southeastern Texas as being expected to be relatively [(*warmer and drier*) (*warmer and wetter*)(*cooler and drier*)(*cooler and wetter*)] than normal.

4. Consider the combination of temperature and precipitation anomalies anticipated for eastern Montana and North Dakota during the same time period. One can characterize eastern Montana and North Dakota as likely to be relatively [(*warmer and drier*) (*warmer and wetter*)(*cooler and drier*)(*cooler and wetter*)] than normal.

5. Based on the temperature and precipitation anomalies anticipated for Alaska, one could expect May/June/July 2011 to be
[(*warmer and drier than normal*)(*warmer and wetter than normal*)
(*cooler and drier than normal*)(*cooler and wetter than normal*)
(*equally likely above or below normal in temperature and/or precipitation*)].

6. Finally, consider the same time period for New England (Maine, New Hampshire, Vermont, Massachusetts, Rhode Island and Connecticut). It was expected to be
[(*warmer than normal but equally likely for above or below normal in precipitation*)
(*cooler than normal but equally likely for above or below normal in precipitation*)
(*equally likely above or below normal in temperature but wetter than normal*)
(*equally likely above or below normal in temperature but drier than normal*)
(*equally likely above or below normal in temperature and/or precipitation*)].

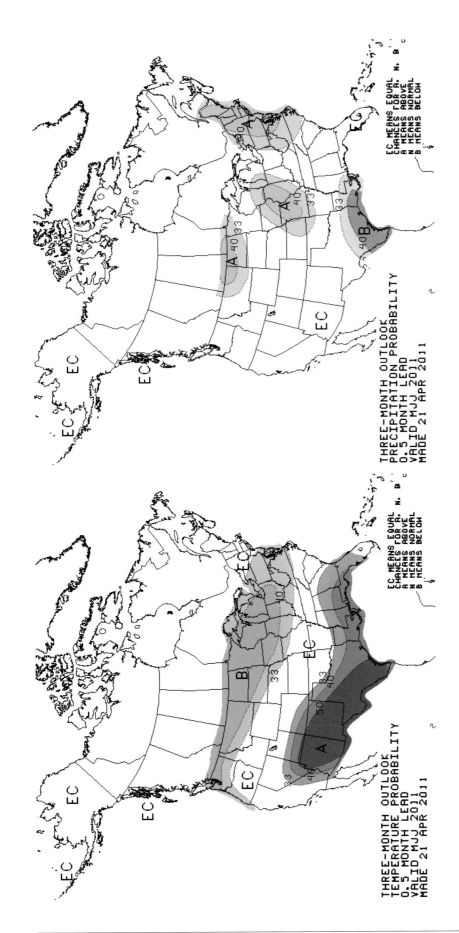

Figure 1.
May-June-July three-month outlook for temperature (left) and precipitation (right) for U.S., issued 21 April 2011.

La Niña Impacts on U.S. Weather: The climate outlooks of NOAA's CPC are based on a number of factors including ENSO, regional climatic trend continuance, 30-60 day tropical oscillation (Madden-Julian oscillation), North Atlantic Oscillation, Pacific Decadal Oscillation, persistently wet/dry soil moisture conditions, statistical forecast tools, and dynamical forecast models. The tropical Pacific statement from the CPC issued on 7 April 2011 stated, "La Niña weakened for the third consecutive month ... La Niña will continue to have global impacts even as the episode weakens through the Northern Hemisphere spring".

The CPC reported on 9 June 2011 that a transition in the tropical Pacific Ocean from La Niña conditions to ENSO-neutral conditions occurred during May 2011. However, the CPC expected lingering La Niña-like atmospheric impacts into the 2011 summer. Therefore, it is likely that the May/June/July 2011 outlook was based largely on the climatic variability signal of La Niña. Precipitation expectations also involved some soil moisture conditions.

Figure 2 from the CPC shows a composite of thirteen years of temperature anomalies and their frequency (top) and precipitation anomalies and frequency (bottom) for the years when La Niña conditions prevailed during May/June/July (MJJ).

The anomalies map to the left on each level shows the average amount of the anomaly while the frequency map to the right displays the percentage of the years when those anomalous conditions existed. In the left temperature anomalies map, yellow and brown shadings indicate positive anomalies (i.e. average temperatures above normal) while blues show negative anomalies. The frequency shadings on the right indicate in yellows and browns occurrences in greater than half of the years while blues show less than half occurrences of that anomaly sign. The precipitation maps are in blue for positive anomalies (wetter than normal) while yellows and browns are negative (drier). The frequency shadings are the same as for the temperatures.

7. As a general pattern characteristic of La Niña during MJJ, the central portion of the U.S. could be described as relatively [(*warm and dry*)(*warm and wet*)(*cool and dry*)(*cool and wet*)].

8. This La Niña MJJ pattern, particularly for western Nebraska and western Kansas, has been noted in [(*less than*)(*more than*)] half of the La Niña years.

9. In Figure 2, areas of relatively strong precipitation deficits were indicated to occur during La Niña years in southern Iowa - northern Missouri and along the Gulf Coast in Texas. The precipitation outlook of Figure 1 suggested that this below normal precipitation during May to July 2011 was more likely to occur in those portions of [(*Iowa*)(*Texas*)].

The latest ENSO statements from CPC can be found at *http://www.cpc.ncep.noaa.gov/products/analysis_monitoring/enso_advisory/index.shtml*.

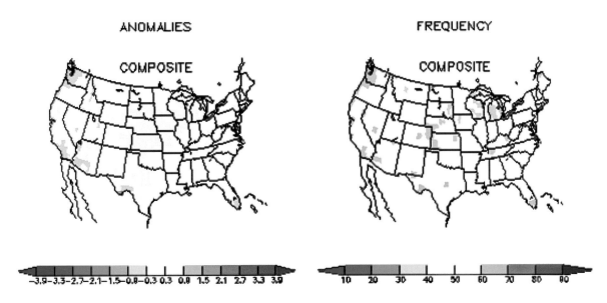

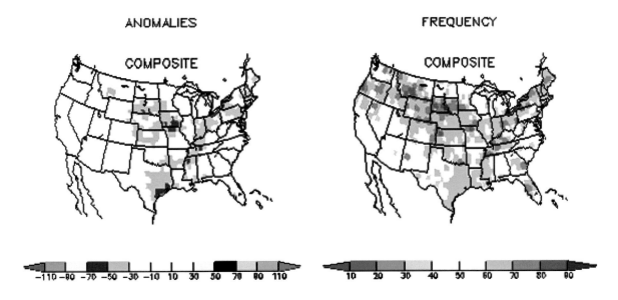

Figure 2.
Early summer season temperature (in degrees Celsius) and precipitation (in millimeters) anomalies and frequencies from thirteen La Niña years.

As knowledge of other climatic variations inherent in Earth's climate system become better understood, short to medium range climate forecasting ("Outlooks") will rely on their signals for information of use to agriculture, energy, and transportation sectors of the economy to minimize impacts and save money. Information regarding these various internal fluctuations:

North Atlantic Oscillation, Pacific Decadal Oscillation, Arctic and Antarctic Oscillations, can be accessed from the course website, under **Climate Variability** links.

Summary: Climatic data provide valuable information for planning purposes. These outlooks result in numerous benefits by enabling many sectors (e.g., agriculture, building industry) to plan ahead for efficient use of resources and taking actions to reduce possible losses.

14A - 1

CLIMATE MITIGATION AND ADAPTATION STRATEGIES

Driving Questions: *What are climate change mitigation and adaptation? What strategies can be applied to lessen the effects of climate change?*

Educational Outcomes: To describe climate change mitigation and adaptation in terms that address the causes of climate change and tackle the effects of climate change. To identify and explore possible mitigation strategies. To recognize that adaptation strategies are essential to reduce the severity and costs of climate change, although adaptation alone will not adequately meet the challenges of climate change. To understand that climate mitigation and adaptation are not alternative strategies; both must be developed and implemented.

Objectives: After completing this investigation, you should be able to:

- Describe climate change mitigation and adaptation.
- Explain mitigation and adaptation strategies for reducing the impacts of climate change within the context of sustainable development.

The Climate Change Challenge

Observational data show that climate change is unequivocal and global warming over the past 50 years is primarily the result of heat-trapping gas emissions due to human activity. This places all of us in the position of making choices at individual, local, regional, national and global levels that will have profound impacts on our Earth environment and on society.

We can ignore or deny that climate change is taking place and follow a "business as usual" path. To do nothing is a choice. It is a choice that will force us to fatalistically accept whatever impacts climate change delivers to us and to future generations. Doing nothing ignores science-based projections that foretell dire consequences for hundreds of millions of people and major impacts on all of Earth's inhabitants.

Or, we can consider and implement options in response to the climate challenge. The two general categories for responding are through mitigation and adaptation. Mitigation addresses the causes of climate change and adaptation tackles the effects. Essentially, mitigation is about reducing heat-trapping gas emissions to prevent dangerous climate change whereas adaptation is coping with those impacts that cannot be avoided.

Mitigation and adaptation are not alternatives. Both are necessary in addressing the climate challenge. The longer the delay in doing both, the greater the magnitude of negative consequences, and the greater the costs. The lack of response to changing conditions has already committed Earth's climate system to change, including some that are irreversible (e.g., species extinction, sea level rise).

The IPCC defines **mitigation** as a human intervention to reduce the sources or enhance the sinks of greenhouse gases. More specifically, mitigation is a human action that (a) <u>reduces</u> any process, activity or mechanism that releases a greenhouse gas, an aerosol or a precursor of a greenhouse gas or aerosol into the atmosphere, or (b) <u>enhances</u> any process, activity or mechanism that removes a greenhouse gas, an aerosol or a precursor of a greenhouse gas or aerosol from the atmosphere.

1. An example of climate change mitigation that fits the (a) part of the IPCC's mitigation definition is [(*<u>decreasing</u>*)(*<u>increasing</u>*)] the amount of gasoline burned in cars and trucks, thereby reducing carbon dioxide emissions.

2. An example of climate change mitigation that fits the (b) part of the IPCC's mitigation definition is [(*<u>decreasing</u>*)(*<u>increasing</u>*)] land areas covered by forests, thereby having the net effect of removing carbon dioxide from the atmosphere.

The IPCC defines **adaptation** as adjustment in natural or human systems in response to actual or expected climatic stimuli or their effects, which moderates harm or exploits beneficial opportunities.

As noted in the IPCC report (2007), two existing U.S. instances involve lands and shorelines where public actions affect change issues. The first is the New Jersey Coastal Blue Acres land acquisition program. Its purpose is to acquire coastal lands damaged or prone to damage by storms or providing buffer for other lands. The second is the common law establishment of a "rolling easement" in Texas. This is an entitlement to public ownership of property that can "roll" inland with the coastline as sea level rises. Other coastal policies exist to encourage coastal landowners to act in ways that anticipate possible sea-level rise.

3. These land acquisition and management programs are examples of climate change [(*<u>mitigation</u>*)(*<u>adaptation</u>*)] strategies.

Climate Change Mitigation Strategies

Identifying and implementing mitigation strategies is a complex challenge. While the goal of ultimately reducing the concentrations of heat-trapping gases in Earth's atmosphere is clear, the path towards its achievement is not. First, the goals of stabilizing and then reducing global emissions must be met. Realization of these goals must occur via complex interactions among environmental, social, and technological processes within the context of broader societal goals such as sustainable development and equity.

Figure 1 identifies by sector the greenhouse gas emissions in the United States (2009). Emissions by sector gives focus to where mitigation efforts need to be directed to be most effective.

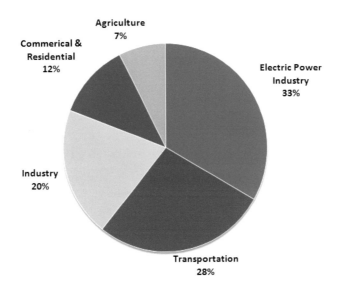

Figure 1.
U.S. greenhouse gas emissions by sector (2009). [Pew Center on Global Climate Change and EPA, Inventory of U.S. Greenhouse Gas Emissions and Sinks: 1990-2009, Table ES-7, 2011]

4. According to Figure 1, it appears that while mitigation efforts should be directed to all sectors, those likely to have the greatest impact on reducing emissions would center on the [(*agricultural and industrial*)(*electrical generation and transportation*)(*industrial and transportation*)] sectors.

One strategic approach to mitigation that relies on current technology has been proposed by Stephen Pacala and Robert Socolow of the Carbon Mitigation Initiative. The principle is that an effective overall policy can be constructed by combining a series of modest actions utilizing current technology. This approach breaks down the mitigation problem into "wedges," with each wedge representing a strategy that can reduce carbon emissions in increasingly greater quantities over time. The stabilization wedges concept is a simple framework for understanding carbon emission cuts that are necessary and the tools already available to make the cuts. This is shown in **Figure 2**. The wedges they propose include (a) building two billion cars with gas mileage of 60 miles per gallon instead of 30 miles per gallon, (b) building two million 1-megawatt wind turbines to displace coal power, (c) doubling nuclear power generation to displace coal power, and (d) capturing and storing greenhouse gas emissions at 800 large coal-fired plants.

5. Figure 2 indicates that current global emissions are about 8 gigatons (billions of tons) of carbon equivalent per year, and the current (business as usual) projected path shows that in 50 years the annual emission is likely to be approximately [(*1.6*)(*8*)(*16*)] gigatons.

6. As drawn, Figure 2 shows that the each of the wedges will be tasked to reduce emissions by [(*1*)(*8*)(*16*)] gigaton(s) per year by the end of the 50-year period.

7. With the wedge approach as shown in Figure 2, the result will be that during the 50-year period the annual amount of actual emissions will [(*increase*)(*decrease*)(*remain the same*)].

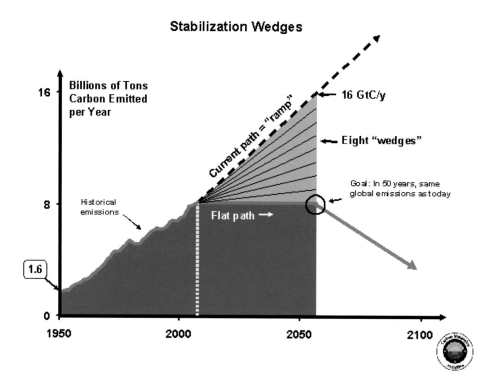

Figure 2.
The "wedge" approach to stabilizing and reducing global emissions. [Carbon Mitigation Initiative, Princeton University]

Even with implementation of the wedge approach, the concentration of carbon dioxide in the atmosphere will exhibit a significant increase over the 50-year period. An associated increase in global average surface temperature can be assumed. To lower temperature will require a decrease in atmospheric CO_2 concentration. Even then, there will be a considerable lag time between lowering CO_2 emissions (as shown in Figure 2) and lowering temperature. However, the necessary first step is to stabilize CO_2 concentrations. The Carbon Mitigation Initiative's stabilization wedges concept shows a practical way to formulate mitigation strategies, at least as a way to lessen the magnitude of climate change.

Regardless of the mitigation strategies that are implemented, it is a certainty that earlier cuts in emissions would have a greater effect in reducing climate change than comparable reductions made later [USGCRP]. A wait-and-see strategy, or simple procrastination, can carry great cost.

Climate Change Adaptation Strategies

As greater atmospheric concentration of heat-trapping gases likely leads to higher surface temperatures, Earth is committed to additional warming no matter what emission mitigation strategies are implemented. Adaptation efforts are essential to reduce the severity and costs of climate change impacts for decades (and perhaps centuries) to come.

Whereas mitigation is more global, adaptation tends more towards regional and local perspectives and actions. Adaptive actions are by individuals or systems in order to avoid, withstand, or take advantage of current and projected climate changes and impacts. Adaptation decreases a system's vulnerability, or increases its resilience to impacts (Pew Center on Global Climate Change).

The USGCRP's *Global Climate Change Impacts in the United States* report (p. 12), available via the course website "Societal Interactions and Climate Policy" section, presents Key Findings that summarize the current and projected climate changes and impacts that require national and local adaptation strategies.

8. According to the USGCRP's listing of Key Findings, the nation's water resources will be stressed by climate change. The geographical area likely to be most impacted by drought associated with climate change will be [(*the Northeast*)(*the Southeast*)(*the West*) (*Alaska*)].

9. According to the USGCRP's listing of Key Findings, coastal areas are at increasing risk from sea-level rise and storm surge, especially along the [*Atlantic and Gulf Coasts*) (*Pacific Islands*)(*parts of Alaska*)(*all of these*)].

On 28 March 2012, the IPCC published a detailed report on *Managing the Risks of Extreme Events and Disasters to Advance Climate Change Adaptation (SREX)*, available at: *http://www.ipcc-wg2.gov/SREX/*. The press release announcing the publication states, "Evidence suggests that climate change has led to changes in climate extremes such as heat waves, record high temperatures and, in many regions, heavy precipitation in the past half century." The report (on page 2 of *Summary for Policymakers*) points out that, "The character and severity of impacts from climate extremes depend not only on the extremes themselves but also on exposure and vulnerability. … Disaster risk management and adaptation to climate change focus on reducing exposure and vulnerability and increasing resilience to the potential adverse impacts of climate extremes, even though risks cannot fully be eliminated."

The SREX report points out that some, but not all, extreme weather and climate lead to disasters. Actions taken to avoid, prepare for, respond to, and recover from the risks of disaster can reduce the impacts of these events and increase the resilience of people caught up in the extreme occurrences. **Figure 3** delineates the SREX adaptation and disaster risk management approaches for facing the extreme events associated with climate change. These approaches can be overlapping and simultaneously pursued.

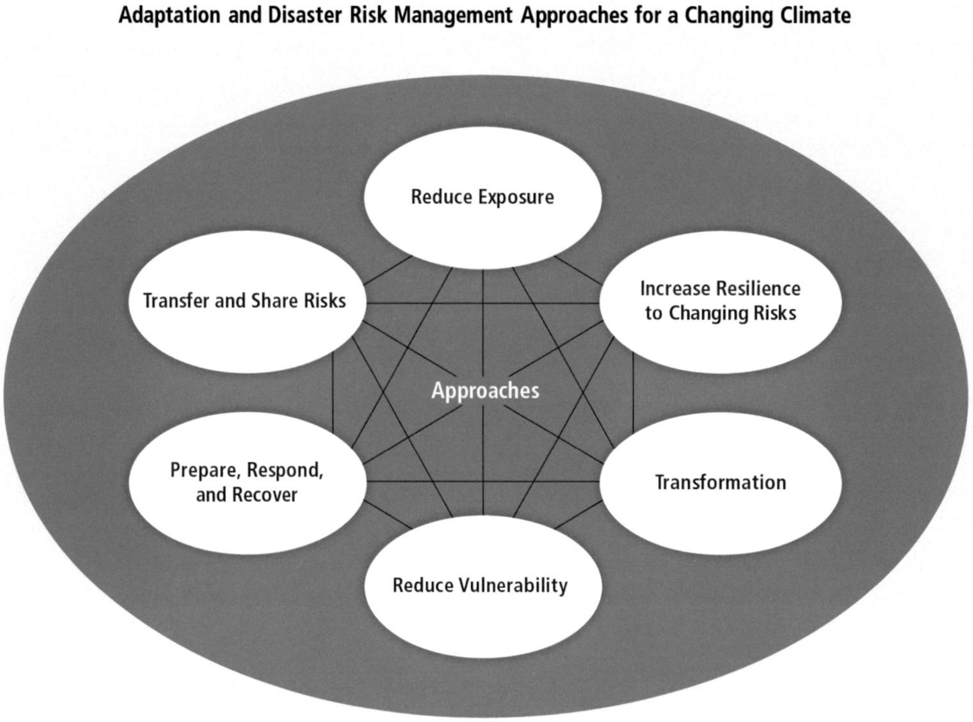

Adaptation and Disaster Risk Management Approaches for a Changing Climate

Figure 3.
Depiction of complementary adaptation and disaster risk management approaches. [IPCC SREX report, Figure SPM.2]

10. The IPCC SREX report assesses the implications of climate extreme events for society and sustainable development. It points out that adaptation and disaster risk management approaches should focus on [(*increasing societal resilience*)(*reducing exposure*)(*reducing vulnerability*)(*all of these*)], even though risks cannot be totally eliminated.

On the global scale, adaptation strategies take on additional significance for developing countries. Good adaptation and good development policy cannot be separated. It is especially important that developed countries provide assistance to these countries as their development and their response to climate change might not be in alignment due to their inadequate resources.

Unfortunately, there are limits to adaptation. Particularly, some developing small island states have few options to adapt to rising sea level. Some ecosystems can be irretrievably lost.

Summary: Adequate response to climate change requires both mitigation and adaptation to lessen impacts. In general, the more mitigation that takes place, the less the impacts to which we will have to adjust through adaptation. Time is of the essence in responding to climate change as some changes will have great impact on hundreds of millions of people and some changes are irreversible.

Investigation 14B: GEOENGINEERING THE CLIMATE

Driving Questions: *What are some of the geoengineering methods that have been proposed to moderate climate change? What is their potential in terms of possible effectiveness, costs, and environmental impacts?*

Educational Outcomes: To define what is meant by geoengineering the climate. To explain proposed geoengineering strategies that reduce atmospheric concentration of carbon dioxide or lower the amount of solar radiation that is absorbed in Earth's climate system. To become aware that geoengineering techniques are as yet unproven and their use is very likely to have associated limitations, uncertainties, and risks.

Objectives: After completing this investigation, you should be able to:

- Describe what is meant by geoengineering the climate, and identify the scientific bases on which geoengineering schemes would operate.
- Present examples of geoengineering-scale carbon dioxide removal and solar radiation management techniques that have been proposed, including their potential, limitations, uncertainties, and risks.

Figure 1.
The original climate conference. [With permission. *http://www.geekculture.com/joyoftech/joyarchives/1332.html*]

Human Impacts on Earth's Climate System

People have been transforming Earth's climate system in increasingly aggressive ways since the beginning of the industrial age and the exponential growth in the global human population. With the burning of fossil fuels, concentrations of atmospheric heat-trapping gases and aerosols have increased. At the same time, about half of Earth's land surface has been reshaped by human activity, thereby impacting the flow of radiant energy to and from our planet. Additionally, numerous other climate mechanisms have been impacted in direct and indirect ways by human activity. The net result has been anthropogenic global warming, with the prospect of greater climate change looming in the future. The likelihood is almost certain of major negative (including catastrophic and irreversible) impacts on human society and Earth's terrestrial and marine ecosystems in the decades and centuries ahead.

Recent science suggests that it may still be possible to avoid the most disastrous impacts of climate change through mitigation and adaptation. However, this requires immediate and decisive efforts to both cut emissions and assist vulnerable countries to adequately adapt. As the results of the December 2009 United Nations Climate Change Conference (COP15) in Copenhagen demonstrate, such immediate and decisive actions appear unlikely. (See **Figure 1** for a depiction of what the first climate conference might have been like.)

Even with a rapid response to climate change, mitigation and adaptation may not be sufficient to prevent dangerous anthropogenic climate change. Additional, not alternative, schemes may be required and merit serious consideration. These additional strategies, planetary in scale, are grouped under the term **geoengineering**. Geoengineering could lessen global climate change, but with the risk of unanticipated consequences.

Anticipating that the response of the world community will be too-little and too-late, some scientists have been exploring more drastic last-ditch actions to cool a runaway climate. They propose serious consideration of physical interventions, or technological fixes, on a global scale. Proposed is **geoengineering** with the ultimate goal of manipulating the global climate by correcting Earth's radiative imbalance with space.

The American Meteorological Society, the professional scientific society representing the atmospheric and related oceanic and hydrologic sciences that form the core of climate and climate change science, has issued ***A Policy Statement on Geoengineering the Climate System***. This has been done in recognition that the time has arrived to address a planetary-scale issue related to manipulating Earth's climate system—a profound issue which was unthinkable until recently. The ***Statement*** follows:

AMS Policy Statement on Geoengineering the Climate System
(Adopted by the AMS Council on 20 July 2009)

Human responsibility for most of the well-documented increase in global average temperatures over the last half century is well established. Further greenhouse gas emissions, particularly of carbon dioxide from the burning of fossil fuels, will almost certainly contribute to additional widespread climate changes that can be expected to cause major negative consequences for most nations[1].

Three proactive strategies could reduce the risks of climate change: 1) mitigation: reducing emissions; 2) adaptation: moderating climate impacts by increasing our capacity to cope with them; and 3) geoengineering: deliberately manipulating physical, chemical, or biological aspects of the Earth system[2]. This policy statement focuses on large-scale efforts to geoengineer the climate system to counteract the consequences of increasing greenhouse gas emissions.

Geoengineering could lower greenhouse gas concentrations, provide options for reducing specific climate impacts, or offer strategies of last resort if abrupt, catastrophic, or otherwise unacceptable climate-change impacts become unavoidable by other means. However, research to date has not determined whether there are large-scale geoengineering approaches that would produce significant benefits, or whether those benefits would substantially outweigh the detriments. Indeed, geoengineering must be viewed with caution because manipulating the Earth system has considerable potential to trigger adverse and unpredictable consequences.

Geoengineering proposals fall into at least three broad categories: 1) reducing the levels of atmospheric greenhouse gases through large-scale manipulations (e.g., ocean fertilization or afforestation using non-native species); 2) exerting a cooling influence on Earth by reflecting sunlight (e.g., putting reflective particles into the atmosphere, putting mirrors in space, increasing surface reflectivity, or altering the amount or characteristics of clouds); and 3) other large-scale manipulations designed to diminish climate change or its impacts (e.g., constructing vertical pipes in the ocean that would increase downward heat transport).

Geoengineering proposals differ widely in their potential to reduce impacts, create new risks, and redistribute risk among nations. Techniques that remove CO_2 directly from the air would confer global benefits but could also create adverse local impacts. Reflecting sunlight would likely reduce Earth's average temperature but could also change global circulation patterns with potentially serious consequences such as changing storm tracks and precipitation patterns. As with inadvertent human-induced climate change, the consequences of reflecting sunlight would almost certainly not be the same for all nations and peoples, thus raising legal, ethical, diplomatic, and national security concerns.

Exploration of geoengineering strategies also creates potential risks. The possibility of quick and seemingly inexpensive geoengineering fixes could distract the public and policy makers from critically needed efforts to reduce greenhouse gas emissions and build society's capacity to deal with unavoidable climate impacts. Developing any new capacity, including geoengineering, requires resources that will possibly be drawn from more productive uses. Geoengineering technologies, once developed, may enable short-sighted and unwise deployment decisions, with potentially serious unforeseen consequences.

Even if reasonably effective and beneficial overall, geoengineering is unlikely to alleviate all of the serious impacts from increasing greenhouse gas emissions. For example, enhancing solar reflection would not diminish the direct effects of elevated CO_2 concentrations such as ocean acidification or changes to the structure and function of biological systems.

Still, the threat of climate change is serious. Mitigation efforts so far have been limited in magnitude, tentative in implementation, and insufficient for slowing climate change enough to avoid potentially serious impacts. Even aggressive mitigation of future emissions cannot avoid dangerous climate changes resulting from past emissions, because elevated atmospheric CO_2 concentrations persist in the atmosphere for a long time. Furthermore, it is unlikely that all of the expected climate-change impacts can be managed through adaptation. Thus, it is prudent to consider geoengineering's potential benefits, to understand its limitations, and to avoid ill-considered deployment.

Therefore, the American Meteorological Society recommends:

1. *Enhanced research on the scientific and technological potential for geoengineering the climate system, including research on intended and unintended environmental responses.*

2. *Coordinated study of historical, ethical, legal, and social implications of geoengineering that integrates international, interdisciplinary, and intergenerational issues and perspectives and includes lessons from past efforts to modify weather and climate.*

3. *Development and analysis of policy options to promote transparency and international cooperation in exploring geoengineering options along with restrictions on reckless efforts to manipulate the climate system.*

Geoengineering will not substitute for either aggressive mitigation or proactive adaptation, but it could contribute to a comprehensive risk management strategy to slow climate change and alleviate some of its negative impacts. The potential to help society cope with climate change and the risks of adverse consequences imply a need for adequate research, appropriate regulation, and transparent deliberation.

---

[1]*For example, impacts are expected to include further global warming, continued sea level rise, greater rainfall intensity, more serious and pervasive droughts, enhanced heat stress episodes, ocean acidification, and the disruption of many biological systems. These impacts will likely lead to the inundation of coastal areas, severe weather, and the loss of ecosystem services, among other major negative consequences.*

[2]*These risk management strategies sometimes overlap and some specific actions are difficult to classify uniquely. To the extent that a geoengineering approach improves society's capacity to cope with changes in the climate system, it could reasonably be considered adaptation. Similarly, geological carbon sequestration is considered by many to be mitigation even though it requires manipulation of the Earth system.*

1. The AMS Statement defines geoengineering as the deliberate manipulation of the [(*physical*)(*chemical*)(*biological*)(*any or all of these*)] aspects of the Earth system.

2. Overall, the Statement describes that at the present time geoengineering of the climate system is [(*proven and low risk*)(*proven and high risk*)(*unproven and high risk*)].

3. The Statement considers the threat of climate change to be serious and unlikely to be managed adequately by mitigation and adaptation. Further, it states that it [(*is*)(*is not*)] "prudent to consider geoengineerings's potential benefits, to understand it limitations, and to avoid ill-considered deployment."

4. Following the listing of three recommendations concerning geoengineering of Earth's climate system, the Statement emphasizes that "Geoengineering will not substitute for either aggressive mitigation or proactive adaption, but it [(*could*)(*could not*)] contribute to a comprehensive risk management strategy to slow climate change and alleviate some of its negative effects."

As stated early in this course, our global climate is fundamentally the story of solar energy received by Earth being absorbed, deflected, stored, transformed, put to work, and eventually emitted back to space. The three fundamental ways in which this energy balance of Earth with space can be perturbed are by: 1) changing the incoming solar radiation; 2) changing the fraction of solar radiation that is reflected by the Earth system; or, 3) altering the longwave (infrared) radiation from the Earth system to space. Geoengineering strategies primarily seek to address on a global scale the perturbations caused by greenhouse gas emissions by either (1) capturing and sequestering heat-trapping carbon out of the atmosphere or (2) through reflecting a small percentage of the solar energy entering Earth's climate system back into space. Respectively, these are called carbon dioxide removal (**CDR**) techniques and solar radiation management (**SRM**) techniques. Examples of CDR and SRM techniques follow.

5. Serious consideration has been given to emulating explosive volcanic eruptions by injecting SO_2 continuously into the stratosphere, producing a sulfuric acid cloud to scatter solar radiation back to space. This is an example of a [(*CDR*)(*SRM*)] geoengineering technique, which would be expected to lower the average global surface temperature.

6. The unintended consequences of cooling the planet by injecting SO_2 into the atmosphere include no impact on the increasing acidification of the ocean by increasing CO_2 concentrations, and the [(*increase*)(*decrease*)] in direct solar radiation for use as solar power.

Based on volcanic eruption studies, the injection of SO_2 into the atmosphere would also produce drought in some parts of the world as well as acid rain which could decimate land and water ecosystems. The strategy does have the advantage of allowing for quick response—added in sufficient quantities it could lower the global temperature within a year or two of deployment. It is also relatively inexpensive compared to other temperature moderating strategies. So inexpensive, in fact, that it could be accomplished by a single country acting alone. This makes it also very dangerous as it could bring about radical and uneven shifts in climate around the world.

7. This sulfur-aerosol injection plan demonstrates a common characteristic of all geoengineering efforts. It has the potential of producing [(*__beneficial__*)(*__damaging__*) (*__both of these__*)] results.

As terrestrial vegetation grows, it removes large quantities of carbon from the atmosphere during photosynthesis. When plants die and decompose, most of the carbon they stored returns to the atmosphere. Under these conditions, the overall growth and decay cycle is essentially neutral in terms of CO_2 emissions into the atmosphere.

8. The biomass resulting from plant growth may be harvested and sequestered as organic material through burial of trees, crop wastes, or as charcoal (biochar). This is an example of a [(*__CDR__*)(*__SRM__*)] geoengineering technique, which would be expected over time to lower the average global surface temperature.

Biochar, as with other forms of charcoal, is created when organic matter decomposes in a low or zero oxygen environment. Once formed, it is resistant to decomposition. It is known from archaeological sites that biochar can remain in soils for hundreds to thousands of years. Studies have shown that significant biomass sequestration is possible in principle, although it would be a partial solution at best. Biochar, however, has the added advantage of making soils more fertile.

IPCC begins assessment of geoengineering proposals: Geoengineering experts met in Lima, Peru in late June 2011 under the auspices of the Intergovernmental Panel on Climate Change (IPCC) to assess proposals for manipulating Earth's climate system to avoid climate disaster. Their singular goal is to assess whether the proposals are sound science.

The understanding of the physical science basis is still limited and the IPCC will address this through three of its working groups preparing the forthcoming Fifth Assessment Report (AR5) on Climate Change to be published in 2013-14. The IPCC expert group emphasizes it is merely comprehensively evaluating the technologies proposed for geoengineering applications, including all their possible impacts. The IPCC will make no recommendations concerning geoengineering.

In advance of the June 2011 meeting, environmental and human rights organizations from 40 countries sent an open letter to the IPCC protesting against the use of geoengineering to change the climate, especially at a time when there is no real progress on mitigation and adaptation.

Summary: Proposed geoengineering techniques are unproven and potentially dangerous. Like medicines, their desirable benefits are likely to be accompanied by undesirable side effects. Yet, they must be considered in the event that actions are needed to allow time for mitigation and adaptation strategies to take hold or because last-ditch options are needed to counteract devastating effects of climate change.

Want to know more about geoengineering the climate? The United Kingdom's Royal Society (the world's oldest scientific academy) has published a comprehensive report of the topic, entitled ***Geoengineering the Climate: Science, Governance and Uncertainty***. The entire report is available online at: *http://royalsociety.org/geoengineeringclimate/*.

Investigation 15A:

CLIMATE MITIGATION THROUGH CARBON EMISSION CAP-AND-TRADE

Driving Question: *How can heat-trapping carbon dioxide emissions be reduced through a cap-and-trade or carbon tax program?*

Educational Outcomes: To describe the projected global warming for different carbon emissions scenarios. To present the choices the world community faces with the associated global temperature increases. To explain the basics of a cap-and-trade system that employs regulation and market forces and the fundamentals of a carbon tax to achieve mitigation goals.

Objectives: After completing this investigation, you should be able to:

- Describe the relationship between global greenhouse gas emissions and predicted average global temperatures during the 21st century.
- Explain the fundamentals of the cap-and-trade system for reducing global greenhouse gas emissions into the atmosphere.
- Provide an overview of the concept of a carbon tax.

Climate Mitigation through Reductions in Greenhouse Gas Emissions

There is overwhelming agreement among climate scientists that current and anticipated global climate changes are primarily the result of increased concentrations of heat-trapping greenhouse gases, mainly carbon dioxide (CO_2), in the atmosphere. It is known that these increases are the result of human activity, much of which involves the burning of fossil fuels (petroleum, coal, and natural gas) for generation of electricity and for transportation. It is of paramount importance that these emissions be reduced in order to prevent a dangerous rise in atmospheric CO_2.

National and international efforts are underway to develop and implement strategies to significantly reduce emissions of CO_2 and other greenhouse gases. Central to the efforts is the achievement of international agreement on the essential elements of a comprehensive global approach to prevent climate disaster. The goal is to acquire long-term commitments and the launching of immediate action to shape our common future and that of generations to come.

Introduction: The IPCC *Climate Change 2007 Fourth Assessment Report (AR4)* projects climate change and its impacts through this century in terms of global greenhouse gas (GHG) emissions and global surface warming for different emissions scenarios. These projections are summarized in **Figure 1**. Different emissions scenarios have been devised to explore their impacts on global temperature. They are grouped into four scenario families (A1, A2, B1, and B2) that explore alternative development pathways, covering a wide range of demographic, economic and technological driving forces and resulting GHG emissions.

The figure's left panel displays emission rates expected with the different scenarios. The panel to the right shows the 20th century global surface warming curve as well as 21st century global surface warming projections under different emissions scenarios.

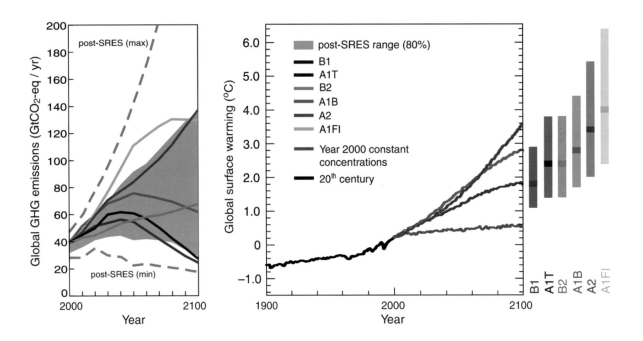

Figure 1.
Scenarios for greenhouse gas (GHG) emissions in the 21st century and projections of surface temperatures. [IPCC, Figure SPM5]

1. In the right panel of Figure 1, the vertical axis is the global surface temperature anomaly (in Celsius degrees) compared to the average temperature during the 1980-1999 time period (i.e., the position of 0 on the temperature anomaly scale). Draw a horizontal line across the graph to represent the 0-degree temperature anomaly. The black curve represents the change in global average surface temperature from 1900 to 2000. It shows that during the 20th century, the global surface temperature increased approximately [(*0.2*) (*0.8*)(*1.2*)] Celsius degree(s).

2. What would happen if the concentration of atmospheric greenhouse gases were to remain constant at their Year 2000 levels? The lowest (magenta) temperature curve extending to the Year 2100 shows that with a steady concentration of greenhouse gases, the global average surface temperature would [(*slightly decrease*)(*remain steady*) (*slightly increase*)] during the 21st century.

3. The blue, green, and red curves in the figure's right panel show best estimates of temperature changes in the other emissions scenarios during the 21st century. They project a 21st century average global temperature warming as much as approximately [(*1.6*)(*2.6*) (*3.4*)] Celsius degree(s).

Even if the global community acts ambitiously together to reduce emissions, the different IPCC scenarios show Earth's climate system is already *committed* to rising temperatures. A business-as-usual (doing-nothing-more) path is potentially disastrous as it is likely to result in substantially greater average global temperature increases in the future. A recent scientific study estimates as much as a 7 Celsius degree rise in temperature is possible by the end of the century. A number of strategies are needed to prevent such warming. One approach of great promise is through a system called cap-and-trade.

Climate Mitigation through Cap-and-Trade

Underlying the **cap-and-trade** approach to reducing greenhouse gas emissions is a simple idea that the marketplace can be a powerful tool for achieving environmental progress. In the program, the **cap** is a legal limit imposed on the quantity of greenhouse gases that a company, region or country can emit each year and **trade** means that companies, regions or countries that emit greenhouse gases may buy and sell the permission, or permits, to emit greenhouse gases up to the overall cap among themselves. The net effects are that as the cap declines over time the total emissions are forced to decrease and, in the trading process, overall costs of abatement are reduced.

Figure 2 is a schematic depiction of how a cap-and-trade system works after it is up and running. To start the system, government at some level sets the target, or cap, on how much pollution (e.g., CO_2) a business sector (or region or country) is allowed to emit into the atmosphere. The cap places a legal limit subject to enforcement on emissions of greenhouse gases by significant polluters. The non-governmental marketplace then is left to figure out how to achieve the target at the lowest possible cost. With commonsense rules set in place, the competitiveness and ingenuity of the marketplace reduces emissions as smoothly, efficiently, and cost-effectively as possible.

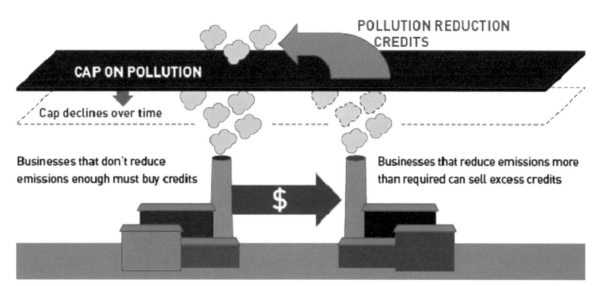

Figure 2.
How an established **cap-and-trade** system works. Cap limits emissions and trade allows trading (selling) of carbon credits. [Copyright © 2009 Environmental Defense. Used by permission.]

After an overall cap has been established, all significant polluters must collectively meet it. In cap-and-trade, businesses that can reduce emissions at relatively low cost do so. They have an incentive to reduce emissions below the amount required as they can profitably sell the differences (called credits) between their emissions and their caps to other businesses. Companies faced with high costs to cut emissions might choose to save money by buying excess pollution credits from other companies rather than paying the costs of pollution abatement.

4. In Figure 2, the company at the right was able to reduce its pollution below its cap. This put the company in the position of being able to sell its excess credits in the marketplace. In the figure, tan puffs denote pollution units and blue puffs depict credit units. It is evident from counting the arbitrary number of puffs in the figure, the company at the right had a cap of 5 pollution units, but it had reduced its pollution to the point that it had [(*2*)(*3*)(*5*)] credits it could sell.

5. In Figure 2, the company at the left did not or could not achieve its pollution cap of 5 pollution units (shown by the 5 tan puffs under the cap) as it was exceeding its emissions cap by [(*2*)(*3*)(*5*)] as shown by the tan puffs above the cap.

6. Both businesses met their legal obligations to maintain pollution concentrations below the combined cap level of ten pollution units. This was possible because the business at the left could buy [(*2*)(*3*)(*5*)] excess credits from the business at the right.

7. Summarizing, Figure 2 shows the flow of pollution reduction credits from companies that reduce emissions below their cap to companies that emit more than their caps in exchange for money. In the trading process, both companies [(*earn or save money*)(*collectively meet the emissions cap*)(*both of these*)].

8. Figure 2 shows the cap on emissions as decreasing over time. This drives companies towards adopting clean technology infrastructure as the diminishing number of pollution reduction credits causes individual credits to become more costly (the concept of supply and demand). This gradual lowering of the cap allows time for the marketplace to find the lowest-cost alternatives to meeting the new emissions caps, minimizing the overall costs of compliance. This stepwise process of lowering the cap guarantees that emissions from these companies will [(*decline*)(*steady out*)(*increase*)] over time.

Cap-and-trade systems are not new. The U.S. Environmental Protection Agency (EPA) claims considerable experience developing and implementing cap-and-trade systems for other environmental problems. For an overview of EPA's cap-and-trade programs, go to: *http://www.epa.gov/captrade/index.html*. There, under Quick Links, click on "Acid Rain Program."

9. The EPA reports that perhaps its most successful cap-and-trade program has been the Acid Rain Program. Its overall goal is to achieve significant environmental and public health benefits through reductions in emissions of sulfur dioxide and [(*nitrogen oxides*)(*ozone*)] — the primary precursors of acid rain.

During the operation of the Acid Rain Program's cap-and-trade system, the atmospheric concentrations of those gases causing acid rain were dramatically reduced. For example the ambient air sulfur dioxide concentration in the U.S. declined 56% between 1980 and 2008.

Climate Mitigation through Carbon Tax:

A **carbon tax** (also called **carbon price**) is an environmental tax that is levied on the carbon content of fuels. The focus of the tax is the carbon dioxide emitted to the atmosphere when fossil fuels (coal, petroleum, and natural gas) are burned. Like cap-and-trade, the purpose of a carbon tax is to control access to the use of the atmosphere as a repository for emitted greenhouse gases. Both approaches (carbon tax and cap-and-trade) forces users to pay for access, and the potential for generating revenue is considerable. How the revenue would be applied has major political implications.

For a thorough review of these proposed incentives that would drive behaviors and investments towards a transition to a low-greenhouse-gas emissions economy, go to the U.S. National Academies National Research Council's 2011 publication *America's Climate Choices* report *Limiting the Magnitude of Future Climate Change* at: *http://www.nap.edu/catalog.php?record_id=12785*. The report concludes that a carbon pricing system (either cap-and-trade, taxes, or a combination of the two) is the most important step for providing needed incentives to reduce emissions.

Summary: Cap-and-trade and carbon tax systems offer market-based approaches to limiting greenhouse gas emissions. But because the operational details are vital to the success or failure of such systems, governments need to experiment and gain experience in order to build the most effective systems possible. The U.S. has already conducted successful cap-and-trade programs, providing ample evidence that cap-and-trade systems could be effective mitigation tools in reducing concentrations of heat-trapping gases in Earth's atmosphere and thereby heading off potentially disastrous climate change.

Investigation 15B: CARBON DIOXIDE EMISSIONS, CARBON FOOTPRINTS, AND PUBLIC POLICY

Driving Questions: *What are some of the individual and national contributions to atmospheric carbon dioxide from energy consumption? How do our American carbon footprints compare to those of other countries? What are American perceptions concerning climate change?*

Educational Outcomes: To describe worldwide carbon dioxide emissions from energy consumption in terms of country and per capita. To compare U.S. contributions of carbon dioxide emissions with those of other countries. To investigate public perceptions concerning the resulting global warming, including those of six audiences, called the *Six Americas*, polled via an audience segmentation analysis.

Objectives: After completing this investigation, you should be able to:

- Describe the national and per capita carbon dioxide emissions from energy consumption for a number of countries around the world.
- Explain carbon footprints while providing an overview of U.S. carbon dioxide emissions compared to other countries.
- Describe U.S. public perceptions concerning climate change according to the *Global Warming's Six Americas Reports*.

Carbon Footprints in a Global Perspective

Observations show that warming of Earth's climate system is unequivocal. It is also the overwhelming consensus of climate scientists that global warming over the past half-century has been due to human-induced emissions of heat-trapping gases. [IPCC, USGCRP] These emissions are bringing about increases in the average global surface temperature while perturbing other components of Earth's climate system.

A measure of the amount of carbon dioxide produced by a person, activity, or nation in a given time interval has been referred to as a **carbon footprint**. Not surprisingly, carbon footprints vary considerably around the world when determined on a national or per capita (per person) basis.

Unlike most ordinary footprints, carbon footprints linger. In fact, the impacts of carbon footprints persist for a very, very long time. The carbon footprints you are creating today will be making their presence known far beyond this century.

It is well established through climate science that increasing concentrations of greenhouse gases, most notably carbon dioxide, due to human activity are largely responsible for observed and projected global climate change. While all scientifically informed persons recognize the human role in the increase in greenhouse gas concentrations in Earth's atmosphere, few have made strong connections between these increases and their own actions.

Table 1 describes the national and per capita carbon dioxide emissions from energy consumption for a selected number of countries for the most recent year of available data (2009) [Energy Information Administration, CIA World Factbook].

Table 1: Carbon Dioxide Emissions from Energy Consumption, 2009

Country	Total CO_2 emissions (Millions of Metric Tons)	Population (Millions)	Per Capita CO_2 emissions (Metric Tons)
United States	5425	307.0	17.7
Canada	541	33.4	16.2
Mexico	444	111.3	4.0
Brazil	420	199.1	2.1
France	397	63.0	6.3
Germany	766	82.4	9.3
United Kingdom	520	62.3	8.4
Russia	1572	140.0	11.2
Iran	527	75.9	6.9
Egypt	192	78.7	2.4
South Africa	450	49.0	9.2
Australia	418	21.3	19.6
China	7711	1322.6	5.8
India	1602	1160.9	1.4
Japan	1098	127.1	8.6
World	30,398	6770.2	4.5

1. According to Table 1, the two countries generating the most total CO_2 emissions from energy consumption in 2009 were China and [(*Russia*)(*India*)(*the United States*)].

2. Of the countries listed in Table 1, the one generating the most CO_2 emissions on a per capita basis was [(*China*)(*the United States*)(*Australia*)].

3. Of the countries listed with populations <u>over 25 million</u>, the one with the greatest per capita CO_2 emissions from energy consumption was [(*China*)(*the United States*)(*Australia*)].

4. The United States reported a 2009 per capita CO_2 emissions rate that was [(*0.25*)(*0.98*)(*3.05*)] times that of China.

5. The United States reported a per capita CO_2 emissions rate that was [(*0.07*)(*4.3*)(*12.6*)] times that of India.

6. With 4.5% of the world population in 2009, the United States was generating about [(*4.3%*)(*18%*)(*32%*)] of the world's total CO_2 emissions from energy consumption.

There are numerous reasons for the broad range of CO_2 emissions by country, including levels of development, transportation needs, heating/cooling requirements, industrial demands, availability of other energy sources, and the efficiency of energy use. Faced with the requirement to reduce global CO_2 emissions, a first step is to analyze the current sources of emissions. Highly detailed data on emissions are compiled by the U.S. Department of Energy at: *http://www.eia.doe.gov/* and by the International Energy Agency at *http://www.iea.org/*.

An Update: The U.S. Energy Information Administration (EIA) annually updates U.S. CO_2 emissions data in August. In August 2011, it reported a 3.5% increase in U.S. energy-related CO_2 emissions during 2010. This was the largest annual percentage increase since 1988. It was attributed primarily to the rebound from the economic downturn experienced in 2008 and 2009. The U.S. Gross Domestic Product (GDP) rose 3.0% in 2010. As of August 2010, the EIA was projecting significantly slower emissions growth over the next decade, averaging 0.2% per year. [*http://www.eia.gov/pressroom/releases/press365.cfm*]

Estimating your Carbon Footprint: To help you in approximating your personal carbon footprint, the U.S. Environmental Protection Agency (EPA) has developed calculators. Go to: *http://www.epa.gov/climatechange/emissions/individual.html*.

There, you can click on "Personal Emissions Calculator" to roughly estimate your personal or family's greenhouse gas emissions and explore the impact of taking various actions to reduce your emissions.

Climate Change and Public Policy: Advances in climate science in recent decades clearly point to the need to develop and implement climate change pubic policy at global, national, regional and local levels with strong mitigation and adaptation components. While some climate change impacts may be positive at local and regional levels, the reality is that most consequences of inaction are very likely to be dire.

Policymaking, in general, requires recognition that a public problem exists. As stated in the course text, policymakers (e.g., elected officials, senior bureaucrats, and other public servants) are more likely to address the issue if it "is easy to understand, is the result of a crisis or catastrophe, has a relatively inexpensive solution, attracts widespread public attention, and spurs demands for action." It is suggested that, as an issue, climate change is not seen by many policy makers (or by the general public) as meeting all of these criteria.

How does the general public view climate change? This is a critical question as the formulation and implementation of a public policy ultimately depends on citizen support. Polling is a primary means for estimating public perceptions and support for public policy that adequately meets the climate change challenge we and our descendents must face in the decades and centuries ahead.

Effective engagement and communication on any public policy issue must start with the recognition that people are diverse and have different psychological, cultural, and political reasons for acting or not acting on a public policy issue. An attempt to identify and better understand how different audiences within the American public respond to the issue of global warming in their own distinct ways has made via national representative surveys conducted by the Yale Project on Climate Change and the George Mason University Center for Climate Change Communication.

The George Mason/Yale effort produced its first study, entitled *Global Warming's Six Americas 2009: An Audience Segmentation Analysis*, based on survey questionnaire responses by 2,129 adults in fall 2008. In-depth measures were made of the public's climate change beliefs, attitudes, risk perceptions, motivations, values, policy preferences, behaviors, and underlying barriers to actions. Six audiences, the *Six Americas*, were distinguishable on these dimensions and exhibited very different levels of engagement with the issue of climate change.

Based on the survey, **Figure 1** shows the Proportion of the U.S. adult population in the *Six Americas*. Labels identifying the six population segments appear in the figure. Detailed descriptions of the segments appear in the published report available online.

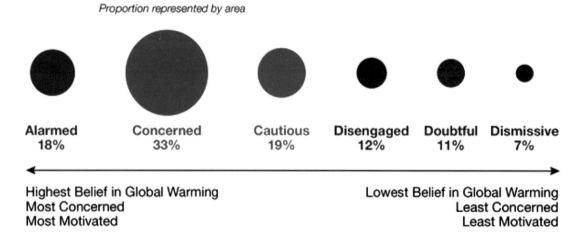

Figure 1.
Proportion of the U.S. adult population in the Six Americas. November 2008 [George Mason/Yale *Global Warming's Six Americas 2009*]

7. Figure 1 reports that the largest identified group, the "Concerned", was composed of [(*7%*)(*11%*)(*18%*)(*33%*)] of the adult U.S. population. Members of this group are convinced global warming is real and serious, but have not yet engaged the issue personally.

8. The "Alarmed" group is already taking action to address the issue. At the opposite end of the spectrum the "Dismissive" group is also actively engaged. The Dismissive group, composed of [(*7%*)(*11%*)(*18%*)(*35%*)] of the adult U.S. population, generally believes

global warming is not happening, is not a threat, and strongly believes it does not warrant a national response.

The Six Americas 2009 report contains highly detailed analyses and is available at: *http://www.climatechangecommunication.org/images/files/GlobalWarmingsSixAmericas2009c.pdf*. For example, if you wish (a) to determine the level of support for a cap and trade policy, go to Figure 22 on page 19, or (b) to evaluate the political party affiliation of the members of the different Six Americas as of fall 2008, go to Figure 29 on page 25.

Update on *Global Warming's Six Americas* Study: The *Global Warming's Six America's nationally representative* survey group was re-assessed in January 2010, June 2010, and May 2011. **Figure 2** provides a summation of the May 2011 survey.

Figure 2.
Proportion of the U.S. adult population in the Six Americas. May 2011. [George Mason/Yale *Global Warming's Six Americas 2011*]

9. Compare Figures 1 and 2. While 51% of those surveyed in November 2008 were alarmed or concerned about climate change, in May 2011 **[(*27%*)(*39%*)(*51%*)]** of those surveyed were alarmed or concerned.

10. The surveys showed that **[(*18%*)(*25%*)(*51%*)]** were doubtful or dismissive about climate warming when surveyed in November 2008, and in May 2011 the total that were doubtful or dismissive of global warming was 25%.

11. The trend of the surveyed group's perceptions about global warming between November 2008 and May 2011 was towards **[(*less*)(*more*)]** concern about climate warming.

This trend was consistent with the levels of awareness of the survey respondents to the increasingly high level of scientific certainty that global climate warming is a reality. On page 7 of the May 2011 report, "Only in the Alarmed and Concerned groups were a majority aware that most scientists think global warming is occurring. Majorities in the other four groups said that either there was a lot of disagreement among scientists or that they didn't know.

Even among the Alarmed and Concerned, however, awareness of the *strength* of scientific agreement is low: While approximately 97% of publishing climate scientists agree that climate change is occurring and that it is caused primarily by human activities, this high level of scientific agreement is recognized by only 44% of the Alarmed, 18% of the Concerned, 12% of the Cautious, and 5% or fewer of the Disengaged, Doubtful and Dismissive."

During 12-30 March 2012, interviews were conducted with 1,008 adults (18+) as part of the ongoing Climate Change in the American Mind series to gather new information on Americans' climate change and energy beliefs, attitudes, policy support, and behavior. Among reports issued from this survey were: *Public Support for Climate and Energy Policies in March 2012* and *Extreme Weather, Climate & Preparedness in the American Mind*. These and other reports of this joint George Mason University/Yale University project can be accessed via: *http://www.climatechangecommunication.org/resources_reports.cfm*.

Table 3 from page 7 of *Public Support for Climate and Energy Policies in March 2012* shows responses to different versions of the question: **Do you think global warming should be a low, medium, high, or very high priority for the President and Congress?**

Table 2. Issue Priority of global warming for the President and Congress

	Mar 2012	Nov 2011	May 2011	Jun 2010	Jan 2010	Nov 2008
Very high	12	12	13	17	13	21
High	28	25	27	27	25	33
Medium	32	33	31	33	31	30
Low	28	30	30	23	31	17

12. Table 2 shows that in March 2012 [(*40%*)(*60%*)(*72%*)] of Americans think that global warming should be a very high, high, or medium priority for the President and Congress.

13. Table 2 also shows that between November 2008 and March 2012, [(*fewer*)(*the same percentage of*)(*more*)] survey respondents felt the President and Congress should place a very high, high, or medium priority on global warming issues.

According to the March 2012 survey, among registered voters, 84% of Democrats, 68% of Independents, and 52% of Republicans think global warming should be a priority. The survey also shows that 91% of Democrats, 77% of Independents, and 70% of Republicans think that overall, protecting the environment either improves economic growth and provides new jobs, or has no effect on economic growth or jobs.

Because climate change is such an important issue, there is a steady flow of surveys conducted by major polling organizations.

Summary: The development and implementation of a U.S. climate change public policy requires ample public support and demand for action. Regardless of the high level of confidence that the climate science community has concerning global climate change, it is critically important in policymaking that the general public perceives the threat of global warming. The use of well designed polls is essential in any public policy development and implementation process, including that dealing with climate change. They provide guidance and reality checks with the public. For example, the polls reported in this investigation indicate substantial public support for action, but trends show less certainty on the part of the public even as the scientific community becomes more certain of the effects of damaging climate change in the decades ahead.

The formulation of a U.S. climate change policy and its implementation are formidable tasks. While science informs, society must determine our actions through enlightened policy making.